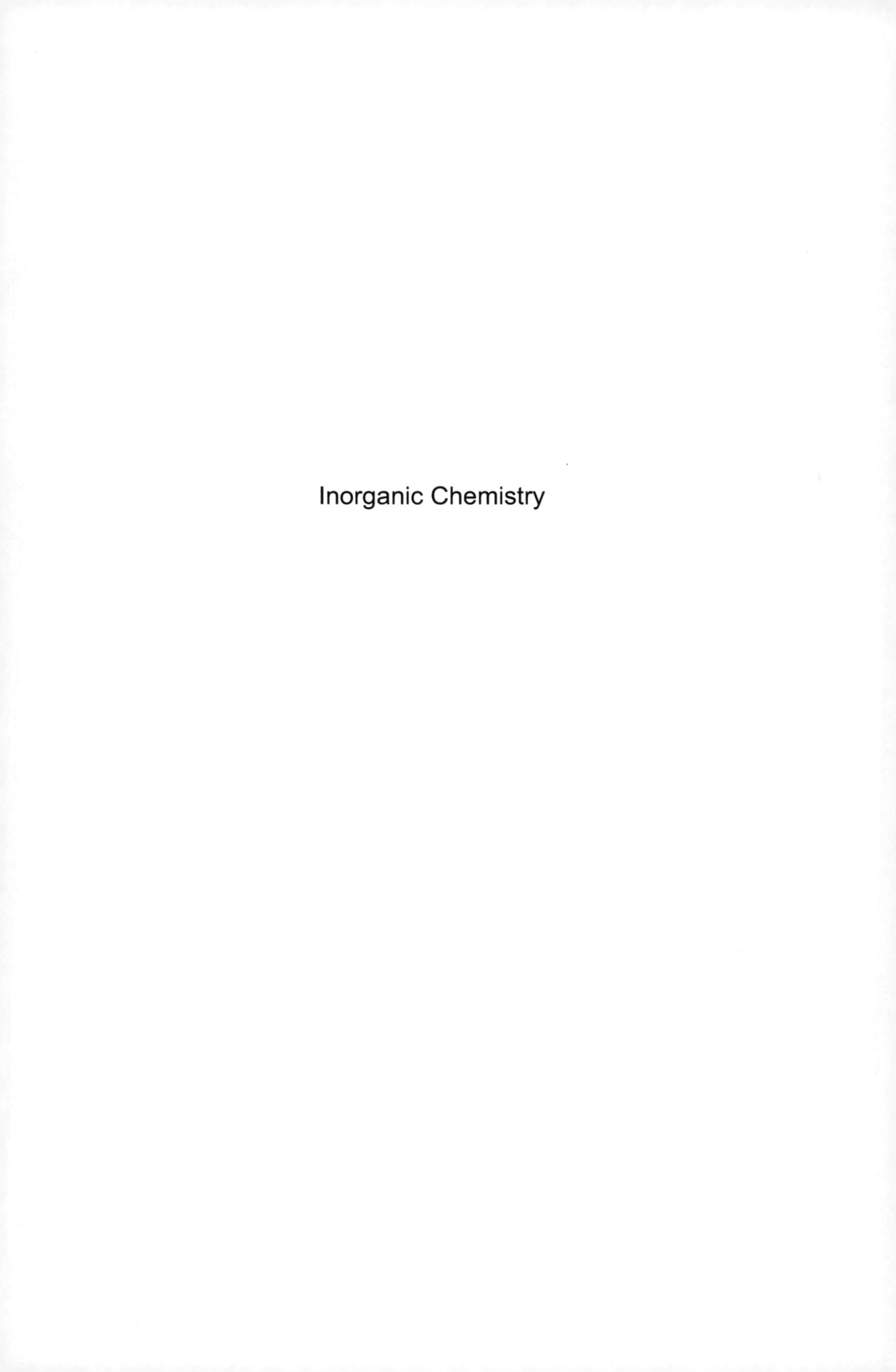

Inorganic Chemistry

Inorganic Chemistry

From Periodic Classification to Crystals

Robert Valls

WILEY

First published 2017 in Great Britain and the United States by ISTE Ltd and John Wiley & Sons, Inc.

Apart from any fair dealing for the purposes of research or private study, or criticism or review, as permitted under the Copyright, Designs and Patents Act 1988, this publication may only be reproduced, stored or transmitted, in any form or by any means, with the prior permission in writing of the publishers, or in the case of reprographic reproduction in accordance with the terms and licenses issued by the CLA. Enquiries concerning reproduction outside these terms should be sent to the publishers at the undermentioned address:

ISTE Ltd
27-37 St George's Road
London SW19 4EU
UK

www.iste.co.uk

John Wiley & Sons, Inc.
111 River Street
Hoboken, NJ 07030
USA

www.wiley.com

© ISTE Ltd 2017

The rights of Robert Valls to be identified as the author of this work have been asserted by him in accordance with the Copyright, Designs and Patents Act 1988.

Library of Congress Control Number: 2017952433

British Library Cataloguing-in-Publication Data
A CIP record for this book is available from the British Library
ISBN 978-1-78630-254-0

Contents

Acknowledgments ix

Introduction xi

Chapter 1. Knowledge of the Periodic Table 1

    1.1. Presentation of the periodic table 1
    1.2. Construction of the periodic table. 2
        1.2.1. History 2
        1.2.2. Structuring of the periodic table 10
        1.2.3. Analysis of various classifications. 14
        1.2.4. Abundance of elements 19
    1.3. Reading the classification 24
        1.3.1. Atomic radius 25
        1.3.2. Electronegativity 28
        1.3.3. Ionization potential 31
        1.3.4. Electron binding energy 34
    1.4. Understanding ions through the classification 37
        1.4.1. The nature and valence of ions through
        the classification 37
        1.4.2. Radius of ions through the classification 41
        1.4.3. Polarizability 44
        1.4.4. The radii of ions in solids 46

Chapter 2. Knowledge of Metallic Crystals. 53

    2.1. Properties of metals 53
        2.1.1. Characteristics of the metallic bond 54
        2.1.2. Conductivity and the melting
        temperature of elements 56

2.2. Study of packing in metals 59
  2.2.1. Formation of planar packing 60
  2.2.2. Crystal formation 62
  2.2.3. Counting atoms in a unit cell 68
  2.2.4. Packing density 71
  2.2.5. Designation of planes in a crystal 73
  2.2.6. Surface density 76
2.3. Representation of metallic crystals 81
  2.3.1. Definition of the unit cell 81
  2.3.2. Geometry of simple polyhedrons 96
  2.3.3. The sites 100
2.4. Packings and diagrams 103
  2.4.1. Reading the diagrams 105
  2.4.2. Solid solutions 109
  2.4.3. Intermetallic compounds 112
  2.4.4. Simple phase diagrams 113

Chapter 3. Knowledge of Ionic Crystals 125

3.1. Description of ionic to covalent crystals 125
3.2. Pauling's rules 129
  3.2.1. The ionic character of a bond
  according to Pauling 130
  3.2.2. Pauling's first rule: coordinated polyhedra 133
  3.2.3. Pauling's second rule: electrostatic valence principle . . 141
  3.2.4. Pauling's third rule: connections of polyhedra 144
  3.2.5. Pauling's fourth rule: separation of cations 146
  3.2.6. Pauling's fifth rule: homogeneity
  of the environment 147
  3.2.7. Presentation of criteria employed 147
3.3. Geometry of binary crystals of MXn type 149
  3.3.1. Presentation of the mentioned compounds 149
  3.3.2. Study of cesium chloride 151
  3.3.3. Study of sodium chloride 159
  3.3.4. Study of zinc sulfide (sphalerite) 171
  3.3.5. Study of zinc sulfide (wurtzite) 178
  3.3.6. Study of nickel arsenide 185
3.4. Geometry of binary crystals of MX_2 type 191
  3.4.1. Study of calcium fluoride 191
  3.4.2. Study of lithium oxide 196
  3.4.3. Study of rutile 199
  3.4.4. Study of cadmium iodide 206
  3.4.5. Study of cadmium chloride 212

3.5. Review of characteristics of binary structures 215
   3.5.1. Crystalline characteristics 215
   3.5.2. Characteristics of availability 216
   3.5.3. Characteristics of the unit cells 217
   3.5.4. Characteristics of the families of compounds 219
3.6. Geometry of ternary crystals of AB_nO_m type 221
   3.6.1. Study of $SrTiO_3$ perovskite 221
   3.6.2. Study of $MgAl_2O_4$ spinel 227

Appendix 237

Bibliography 239

Index 255

Acknowledgments

This book has seen the light of day thanks to many students, colleagues and friends who have directly or indirectly contributed to ideas, asked questions and signaled contradictions. But it is within the family that I found the necessary energy, so I particularly thank my wife, Véronique, for her unwavering support throughout the elaboration of this book, her meticulous proofreading and proposals that proved to be always correct and accurate.

I am deeply grateful to my students, who have contributed to my shaping as a chemist as much as I have contributed to their education, and also to my colleagues, teachers and researchers who have always been available for substantive discussions and also for details, and who have helped me with their suggestions, consistently constructive criticism and advice.

I thank Daniel Farran, for his help and concise and clear remarks, as well as for our pleasant and effective discussions.

Many colleagues, students and friends have participated in bringing this book to light. I sincerely thank all of them.

I can only praise VESTA software, without which the project of this book would have been impossible [MOM 11]; as, in my opinion, the quality of representations is essential for clear presentation of structures. Indeed, not everyone has the same 3D visualization performance, therefore rich and clear representation can aid spatial visualization. It is my intention to offer even those who struggle with volume visualization an opportunity to appreciate this field of chemistry.

Those who want to discover or improve their skills in the field are encouraged to use this simple software for a very rapid construction of their own figures, distance measurement, removal of atoms that obstruct the view, exploration of the solid, etc.

The chemist's reference book, Handbook [HAY 15], has been a valuable source of consistent data that can be easily verified by everyone. Many other references have been used and cited, but the fact remains that the Handbook has served as a hub for the choice of data.

Finally, I should mention *Crystallography Open Database* (COD) as a reference in my knowledge-sharing endeavor, thanks to which I could once again offer consistent and easily verifiable data. It is the source of all the examples of simple compounds cited in this book, including the several atypical compounds that illustrate specific properties.

This book aims at making inorganic chemistry known, facilitating its in-depth knowledge or simply its appreciation by means of the periodic table directed toward knowledge of crystals. Above all, I hope to convince the reader to take a different look at this field of chemistry, which is probably not as well-known as it deserves, all the more so as its applications are very frequent and present in our daily life.

Readers who are willing to contribute to the improvement of this book are invited to send me their critical comments and suggestions. I thank them in advance, as I have always received constructive proposals because of which I was able to simplify, clarify or enhance my published works.

Introduction

An Approach to Inorganic Chemistry

Inorganic chemistry mainly involves the study of objects of the mineral kingdom, as opposed to organic chemistry, which obviously deals with organic compounds. Despite the privative prefix seemingly defining as inorganic that which does not retain what is organic, there is no clear demarcation between these two fields of chemistry. As many compounds fall at the boundary between the two fields, there is no clear-cut division between them.

This division can be illustrated as follows: galleries dug by science in the knowledge reservoir throughout its decades-long progress have followed various lodes, and have come to converge despite their starting point being located in distinct areas of knowledge. This may be the characteristic of various branches of chemistry that have converged over time.

According to the most general definition, inorganic chemistry relates to all non-organic compounds formed of elements that compose the periodic table.

One part of this domain may be represented as a list of structures and their corresponding properties, as well as their chemical reactivity, applications and methods used to obtain the elements and their simplest or most complex derivatives.

With this perspective in mind, a presentation can be elaborated depending on the interest that the compounds present for the branch of chemistry practiced by the author. Moreover, presenting all these elements and everything related to them is always a challenge, since, on the one hand, it is difficult to elaborate an exhaustive presentation, which would come down to plagiarizing encyclopedias, and, on the other hand, it is equally difficult to draw clear limits.

Many books offer these rich and exciting descriptions that make us love chemistry [ANG 07], but this type of description is beyond the scope of the present book.

Given the availability of reliable and comprehensive databases, such as the *Handbook of Chemistry and Physics* [HAY 15], which should always be within our reach, this type of inventory will be omitted here. Moreover, for consistency reasons, that particular work has been used as a source for most of the data in the present book.

Since one of the objectives of this book is to propose easy-to-find and reliable data, links to well-known sites and easy-to-consult files, all accessible to the scientific community, will be provided.

VESTA (*Visualization for Electronic and Structural Analysis*) software has been used for the design of this book, since high quality representation and drawing of the most representative figures may prove to be difficult in the absence of a high performance tool [MOM 11].

All the structures referred to in this book have been taken from Crystallography Open Database (http://www.crystallography.net/), a comprehensive resource featuring over 376 000 entries. A number of these structures can be found in the Handbook [HAY 15], in various books or publications, or in other databases that have been consulted for data consolidation purposes.

A brief history reveals that the most ancient branch of inorganic chemistry is unquestionably metallurgy, whose practice dates back to 2500 BC, and is attributed to Egyptians. Indeed, they practiced mining while Europe was in the Bronze Age and Europe's inhabitants began ore mining and processing, making copper and tin alloys.

The Bronze Age was an essential period marked by the use of metals and, above all, by the development of metallurgy and thus the development of the techniques needed for obtaining bronze (an alloy of copper and tin).

Metallurgy is the study of metals: their extraction, properties and processing. By extension, it is also the name given to the industry producing metals and alloys, which relies on the mastering of this science. This emerging metallurgy required expertise in the art of fire, which was acquired by pottery firing. Extraction of metal from ores depends on the mastering of high-temperature furnaces as copper melts at 1085°C, though the melting point can be significantly reduced by the addition of tin. Subsequent to copper metallurgy, iron metallurgy requires a higher temperature as iron melts at 1538°C, which explains the chronology of the bronze and iron ages.

The wealth of empirical knowledge acquired and transmitted gave rise to alchemy, whose fantasies, still present in the collective unconscious, are often associated with chemistry. The best example is the philosopher's stone, which allows for metal transmutation (turning lead into gold, obviously), but alchemists have gone even further and have imagined the panacea (universal remedy), as well as the elixir of life, which nowadays has taken the more mundane name of brandy.

Chemistry as we know it started with the creation of the periodic table, and Lavoisier can be considered its founding father, as he is one of the first experimental chemists.

Referring to the practice and definition of chemistry, Lavoisier stated the following:

> "As the usefulness and accuracy of chemistry depends entirely upon the determination of the weights of the ingredients and products both before and after experiments, too much precision cannot be employed in this part of the subject; and, for this purpose, we must be provided with good instruments. As we are often obliged, in chemical processes, to ascertain, within a grain or less, the tare or weight of large and heavy instruments, we must have beams made with peculiar niceness by accurate workmen... I have three sets, of different sizes...
>
> The principle object of chemical experiments is to decompose natural bodies, so as separately to examine the different substances which enter into their composition." (Antoine-Laurent de Lavoisier, *Elements of Chemistry*, 1789)

These several phrases render both the essential concept, the fact that chemistry is first of all an experimental science, and the fundamental concept, the fact that chemical substances can be decomposed into simple substances. These two notions will lead to the elaboration of the periodic table.

The periodic table has not historically allowed for an explanation of the properties of elements, since it has been in existence for one century, should Moseley be considered the scientist who completed the work. The reverse has actually been the case, as experimentally determined properties of the elements have helped the community of chemists in the 17th Century in the step-by-step building of this currently omnipresent classification.

Inorganic chemistry is the intersection of many branches of chemistry, and this field often leads to crystals and to an understanding of their packing, which may be more or less complex, but is always surprising. Moreover, the set of elements presented in this book should facilitate the comprehension of the structure of perfect crystals (defects will not be covered).

It is thanks to the use of the periodic table, knowledge of the nature of bonds, notions of symmetry and comprehension of binary phase diagrams that crystals are perceived as less enigmatic.

The purpose of this book is to empirically present, based on observations, the structures of the main crystal compounds and to rely on the periodic table for a better comprehension of their structure, linking it whenever possible to the properties of these crystals.

We have therefore opted for a structure in which chapters can be read randomly, the only guide being a need to access data, a desire to immerse oneself once again in a field of knowledge or the opportunity to compare one's knowledge with that presented in the book. This information will be rooted in observations and will be presented in a more empirical rather than theoretical manner, being accompanied by examples and representations that help in visualizing the properties described.

The first chapter presents a reading of the periodic table that relies on the characteristics of atoms and ions (electronegativity, ionization potential, electron binding energy, etc.). It should contribute to an overall original perspective on the periodic table by providing simple and efficient reading keys.

The second chapter develops the bases of crystallography starting from empirical observations and analyses of the structure of metals. It closes with a description of binary phase diagrams that allows the association of structures with properties and offers basic knowledge of metallurgy.

The third chapter offers a presentation of typical crystals depending on the complexity of their chemical formula and of their packing. The study focuses on MX-type binary crystals (CsCl, NaCl, ZnS sphalerite and wurtzite and NiAs) and then on MX_2-type crystals ($CaCl_2$, Li_2O, TiO_2, CdI_2 and $CdCl_2$) by means of representations offering descriptions in a simple manner. Finally, it presents ternary perovskite crystals ($SrTiO_3$) and spinel ($MgAl_2O_4$).

Visual memory is strongly involved and notions such as ionicity or space availability facilitate a simple and effective approach to the structure and nature of bonds in crystals.

1

Knowledge of the Periodic Table

1.1. Presentation of the periodic table

The periodic table is one of the chemist's everyday tools for retrieving the properties of an element based on its position in this table and the various values associated with it. However, before proceeding to use it, knowledge of several simple keys and of their limits should be acquired and their operation should be mastered.

This table should not be perceived as a visionary anticipation or thought of as being received by a scientist through revelation. It is the result of lengthy and patient efforts sustained by Mendeleev and all the chemists in his time, drawing on the empirical knowledge accumulated throughout the 18th and 19th Centuries by several generations of chemists engaged in the pursuit of this classification, which was supposed to name and organize all the elements. Dozens of scientists should be cited, but in order to keep the presentation simple, only three of them will be mentioned here, namely those whose interventions were essential. The first one was Lavoisier, who in 1789 had formalized the notion of elements, then came Mendeleev, who is recognized as the creator of the periodic table according to his work published in 1869, and finally Moseley, who in 1913 offered the key to the complete form of the periodic table.

The use of the table in the absence of several reading keys may prove discouraging, as it requires great memory and leaves the impression of a series of successive specific cases. Moreover, the method used by chemists in order to memorize the periodic table should differ from the one employed for multiplication tables (it is less mechanical), similar to that applied for memorizing the names, location and characteristics of places encountered every day and which are part of real-life experience, therefore involving affective rather than mechanical memorization.

Resorting to deductive presentations of the periodic table does not simplify its acquisition, since the table, though following nearly mathematical laws for some characteristics of the elements, represents a synthesis of all the diversity of nature when it comes to other characteristics.

1.2. Construction of the periodic table

1.2.1. *History*

In order to allow for a better grasp of the subject, a brief history of the periodic table will be presented, covering five periods that are in our opinion decisive for its elaboration, and whose landmarks are given by the following three important names: Lavoisier, Mendeleev and Moseley.

In the period from Antiquity to the Renaissance, seven metals were known: gold, silver, mercury, copper, iron, lead and tin, as well as several other elements such as sulfur, antimony, arsenic, carbon and phosphorus. By the end of the 17th Century, only these 12 simple substances had been discovered.

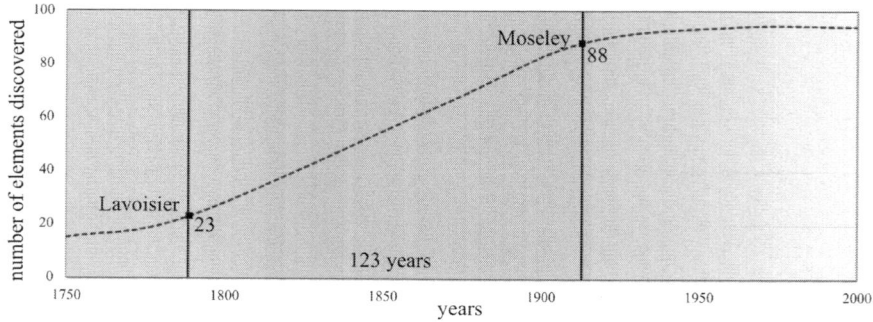

Figure 1.1. *Discovery of elements throughout the centuries*

As the techniques used for analysis evolved, Lavoisier had 23 elements at the beginning of his research work, but their number rapidly increased, so that in 1869 Mendeleev gathered 65 elements in an arrangement that he proposed as the first classification, and then in 1913 Moseley gave the final form of the current periodic table, which contained 88 elements (see Figure 1.1).

1.2.1.1. *The pioneer: Antoine-Laurent de Lavoisier*

In 1789 Lavoisier (1743–1794), with his *Elements of chemistry*, revolutionized (an action fit for those times) chemistry, being the first to propose a distinction between simple and compound substances. He also advanced the idea that certain chemical products, which cannot be decomposed, should be considered elements. It is safe to consider him the founding father of modern chemistry, given his systematic use of weighing instruments. Moreover, he set the bases of chemistry in the following terms:

> "To solve these two questions, it is necessary to be previously acquainted with the analysis of the fermentable substance, and of the products of the fermentation. We may lay it down as an incontestable axiom, that, in all the operations of art and nature, nothing is created; an equal quantity of matter exists both before and after the experiment; the quality and quantity of the elements remain precisely the same; and nothing takes place beyond changes and modifications in the combination of these elements. Upon this principle the whole art of performing chemical experiments depends: We must always suppose an exact equality between the elements of the body examined and those of the products of its analysis." (Antoine-Laurent de Lavoisier, *Elements of Chemistry*, 1789)

A very reductionist but quite effective variant of the above fragment is often stated as follows: *Nothing is created, nothing is lost, everything is transformed*. Lavoisier proposed the idea of simple substances (elements), while he was still far from the periodic table, but he gradually created a ranking (see Table 1.1).

Substances	Elements or assimilated
5 simple substances	Light, caloric, oxygen, nitrogen and hydrogen
6 simple nonmetallic substances	Sulfur, phosphorus, carbon muriatic, fluoric and boracic radicals
17 simple metallic substances	From antimony to zinc, in alphabetical order
5 simple salifiable earth substances	Lime, magnesia, barite, alumina and silica

Table 1.1. *List of simple substances proposed by Lavoisier*

According to Lavoisier's classification, *light* and *caloric* are placed at the same level as oxygen, nitrogen and hydrogen, and this will influence mineral chemistry, which will present the elements in relation with their reactivity toward oxygen, hydrogen and water. Moreover, previous chemical classifications relied on the behavior of substances, their relations or their affinities. Nevertheless, as nothing is perfect, he placed lime, magnesia, barite, alumina and silica in the category of *simple substances*, which he qualified as *salifiable*, and he defined *muriatic, fluoric* and *boracic* radicals without identifying chlorine, fluorine and boron elements.

The table in Figure 1.2 was obtained by placing these elements in the current grid of the periodic table. As the figure shows, a considerable amount of thinking time would have been required for Lavoisier to complete this table, in which the elements he identified in the form of oxide have been indicated by question marks.

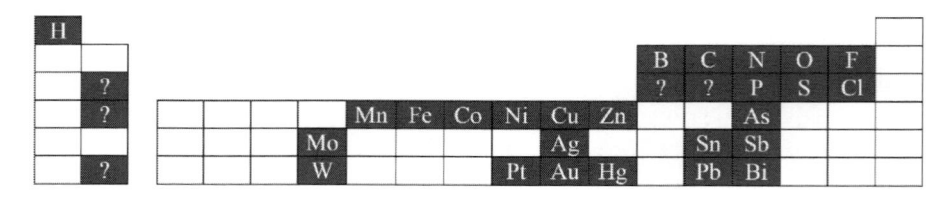

H																	
												B	C	N	O	F	
?												?	?	P	S	Cl	
?				Mn	Fe	Co	Ni	Cu	Zn					As			
				Mo				Ag						Sn	Sb		
?				W				Pt	Au	Hg				Pb	Bi		

Figure 1.2. *First elements available to Lavoisier*

Lavoisier and Guyton Morveau had therefore at their disposal the first 23 elements indicated in Figure 1.2, which are the starting point for the quest to discover new natural elements and classify them, an endeavor in which many scientists have participated.

1.2.1.2. *The creator: Dmitri Ivanovich Mendeleev*

Eighty years later, in 1869, in his paper entitled *On the correlation of the Properties and Atomic Weights of the Elements*, Mendeleev (1834–1907) set the bases of the current periodic table.

He proposed 42 additional elements, reaching therefore a total of 65 available elements and predicted the existence of missing elements (eka-boron, eka-silicon, eka-aluminum and eka-manganese) together with some of their properties.

On 6 March 1869, Mendeleev presented his work at the meeting of the Russian Chemical Society. The following paragraphs of this presentation are, in our opinion, essential:

> "Elements which are similar as regards their chemical properties have atomic weights which are either of nearly the same value (e.g. Pt, Ir, Os) or which increase regularly (e.g. K, Rb, Cs).
>
> The arrangement of the elements or of groups of elements in the order of their atomic weights corresponds with their so-called valences.
>
> The elements which are the most widely distributed in nature have small atomic weights, and all the elements of small atomic weight are characterized by sharply defined properties. They are therefore typical elements. The magnitude of the atomic weight determines the character of an element.
>
> The discovery of many as yet unknown elements may be expected. For instance, elements analogous to aluminum and silicon, whose atomic weights would be between 65 and 75.

The atomic weight of an element may sometimes be corrected by the aid of knowledge of those of the adjacent elements. Thus the combining weight of tellurium must lie between 123 and 126, and cannot be 128. Certain characteristic properties of the elements can be foretold from their atomic weights." (Mendeleev, *On the correlation of the Properties and Atomic Weights of the Elements*, 1869)

The proposal advanced by Mendeleev to use the atomic weight as classification criteria is a first classification key and had the merit of simplicity, as atomic weights were easily measured. Figure 1.3 shows the list of elements classified by Mendeleev located in the grid of the current periodic table.

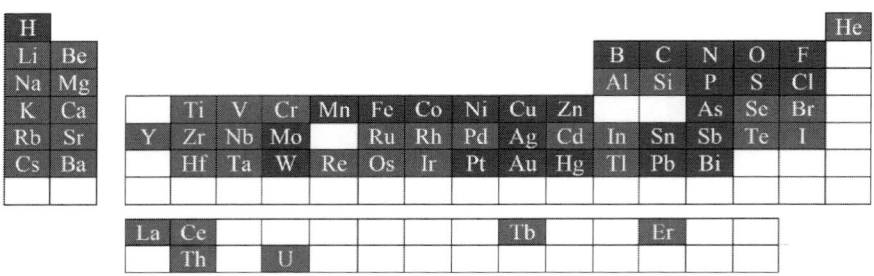

Figure 1.3. *Elements available to Mendeleev. For a color version of this figure, see www.iste.co.uk/valls/inorganic.zip*

The few inversions, cobalt/nickel, argon/potassium and tellurium/iodine (corrected in this table), should not have constituted reasons for doubting this classification proposal, but scientific theories have little tolerance for exceptions unless they are perfectly justified. In the absence of knowledge of the nucleus, no justification was available at that moment.

1.2.1.3. *The finisher: Henry Gwyn Jeffreys Moseley*

A period of 44 years passed until Moseley (1887–1915), while conducting research in Ernest Rutherford's laboratory, established in 1913 the order of elements in the periodic table based on the study of atomic spectra. His work revealed the atomic number as a measurable property and therefore an irrefutable key to the classification of elements.

He evidenced a strong correlation between the X-ray spectrum (see Figure 1.4) of the elements and their location in the periodic table. It was the confirmation of the modern classification, and, considering the 23 additional elements, a number of 88 elements were reached.

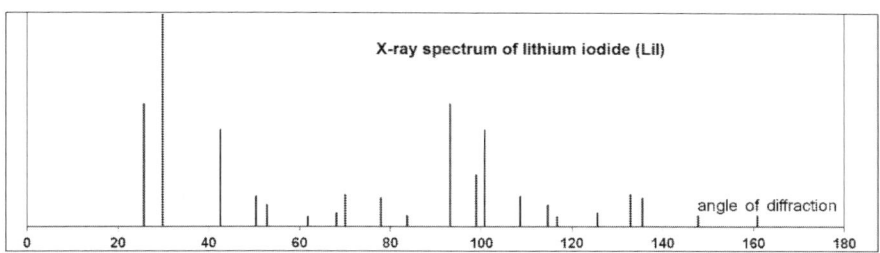

Figure 1.4. *X-ray spectrum of lithium iodide*

To summarize, the first 23 elements were proposed by Lavoisier, then 124 years later the last 23 elements were added, which led in the end to 88 natural elements, which was completed by the last 6 elements (promethium, technetium, francium, astatine, neptunium and plutonium), which are radionuclides produced in laboratories before having been discovered in nature.

The classification was complete and the position of elements was clearly defined; therefore by 1940 the journey was considered completed, after the synthesis of astatine, neptunium and plutonium.

A last development was triggered due to promethium, which had not really been produced and characterized until 1945.

Ninety-four elements have by now been identified in the Earth's crust, some of which have a history that is difficult to summarize, as information on whether the element has been used, predicted, identified, isolated, synthesized or discovered in nature should be provided, as in the case of technetium (43rd element), which has generated considerable interest and whose history is detailed as a note in the paragraph below.

NOTE 1.1.– The difficulty entailed by summarizing the history of elements can be illustrated by a four-stage description of the discovery of element 43:

1) Mendeleev predicted the properties of element 43, which he named eka-manganese. Its place in the classification between molybdenum and ruthenium suggested that it would be an easy-to-discover element.

2) It was erroneously supposed to be present in a platinum alloy in 1828; it was identified in 1846 (but it proved to be niobium), then in 1877 (a mixture of iridium, rhodium and iron), in 1896 (but it was yttrium), in 1908 (it was rhenium), and in 1925 its existence was effectively proved, but the experiment could not be replicated.

The discovery of technetium (43) was finally attributed to Carlo Perrier and Emilio Segrè who isolated the technetium-97 isotope in 1937 (as its name indicates, it was the first element to have been artificially produced).

3) Technetium was detected in red giants in 1952, and its discovery proved that stars could produce heavy elements during nucleosynthesis.

4) On Earth, technetium was isolated in the form of traces in the uranium ore in 1962 [KEN 62].

The difficulty raised by associating a date to this 43rd element is clear, since it was successively predicted, produced, detected and isolated from the natural environment.

The final, though temporary, point of this quest was perhaps the discovery [KEN 62] in the pitchblende (uranium ore) of traces of neptunium, plutonium and technetium.

Since 94 natural elements were isolated in the Earth's crust, there is no necessity to explain the absence of technetium on Earth, and it can be concluded that all natural elements have been obtained.

Resuming the numbering of Newlands' groups [NEW 65] with Roman numerals from I to VIII (law of octaves), Moseley introduced letters A and B, which have been used until the 20th Century. Columns are currently assigned numbers from one to 18, and Roman numerals have been forgotten (though they retain didactic utility when drawing analogies between columns 1 and 2, on the one hand, and columns 11 and 12 on the other hand, or between 3 and 4 and 13 and 14) and the letters attributed to periods were replaced by figures from 1 to 7 (see Figure 1.5).

There are yet several points to choose in the classification in order to adapt it to each one's presentation (see Figure 1.6):

– the position of lanthanum and actinium, and therefore the positioning of the f-block, has been solved by bringing the group of lanthanides and actinides to 15, but lanthanum and lutetium, as well as actinium and lawrencium, can be permutated along the classifications due to their exception to the Madelung rule when filling the f-subshell. In this paper, the atomic number is privileged; therefore lanthanum, the 57th element occupies the 57th place and actinium occupies the 89th place;

– metalloids are always difficult to define, since these elements have at least one metal-like property and one non-metal-like property, while many elements (including non-metals) exist in various crystal forms, each of which has different properties. Moreover, the metalloid, metal or non-metal nature of polonium and astatine, for example, varies depending on the author, with a tendency toward considering polonium a poor metal and astatine the last of halogens.

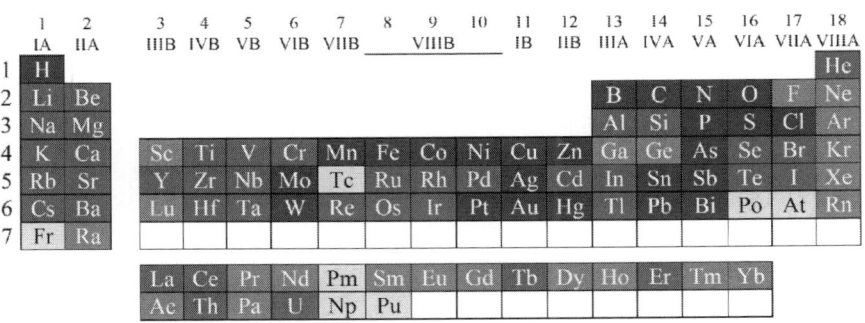

Figure 1.5. *Periodic table according to Moseley. For a color version of this figure, see www.iste.co.uk/valls/inorganic.zip*

The latter point is understandable, given that, besides the fact that the two natural elements are rare, the toxicity exhibited by polonium when in excess of several micrograms, as well as the approximately eight-hour half-life of the most stable isotope of astatine, do not facilitate their manipulation.

Science does not look favorably upon quests that come to an end; so once the research on natural elements has been completed, scientists have turned their attention to synthetic elements.

The Second World War triggered the quest for mastering nuclear forces and opened a new horizon for science. It was the beginning of a new quest for the discovery of heavy or super-heavy elements, an exciting endeavor punctuated by atypical events, which will not be covered in this book.

H																	He
Li	Be											B	C	N	O	F	Ne
Na	Mg											Al	Si	P	S	Cl	Ar
K	Ca	Sc	Ti	V	Cr	Mn	Fe	Co	Ni	Cu	Zn	Ga	Ge	As	Se	Br	Kr
Rb	Sr	Y	Zr	Nb	Mo	Tc	Ru	Rh	Pd	Ag	Cd	In	Sn	Sb	Te	I	Xe
Cs	Ba	Lu	Hf	Ta	W	Re	Os	Ir	Pt	Au	Hg	Tl	Pb	Bi	Po	At	Rn
Fr	Ra	Lr	Rf	Db	Sg	Bh	Hs	Mt	Ds	Rg	Cn	Nh	Fl	Mc	Lv	Ts	Og

		La	Ce	Pr	Nd	Pm	Sm	Eu	Gd	Tb	Dy	Ho	Er	Tm	Yb
		Ac	Th	Pa	U	Np	Pu	Am	Cm	Bk	Cf	Es	Fm	Md	No

Figure 1.6. *Periodic table highlighting several specific elements. For a color version of this figure, see www.iste.co.uk/valls/inorganic.zip*

1.2.2. Structuring of the periodic table

It is important to have in mind a simple and quick description that offers an overall definition of the classification.

The whole periodic table could be summarized as follows:

– sixty-two metals (poor metals not included), which can be listed in descending order of abundance: iron (4th), calcium (5th), sodium (6th; obsolete *natrium*), magnesium (7th), potassium (8th; obsolete *kalium*) and titanium (9th);

– eighteen non-metals, including the noble gases, among which is the most abundant element in the Earth's crust, oxygen, followed in descending order by phosphorus (10th), fluorine, sulfur, chlorine, carbon, hydrogen and nitrogen. As for bromine, iodine, selenium and astatine, these are trace elements with concentrations below 10 ppm. This explains why nitrogen addition has an impact on cultures (its natural presence being scarce) and especially why nitrate pollution has generalized as a consequence of intensive agriculture and the massive addition of nitrates to soils. The six noble gases (previously known as inert or rare gases) are worth a dedicated paragraph, as their designation has evolved in line with our knowledge of the universe. When it was known that helium represented approximately one quarter of the baryonic matter in the universe and argon accounted for about 1% of the terrestrial atmosphere, rare proved no longer suitable as a term describing these gases. As for inert, this term became improper when chemistry developed and hundreds of compounds, the most important being xenon, started being published [BAR 62] in 1962;

– eight metals of the p-block called poor metals, including the third element of the Earth's crust, aluminum, as well as gallium and lead. The rarer ones (below 10 ppm) are tin (obsolete *stannum*), thallium, indium, bismuth and polonium (see Figure 1.7);

– six metalloid elements, including the second element of the Earth's crust, silicon; boron, germanium, arsenic, antimony (obsolete *stibium*) and tellurium are rarer and their concentration is below 10 ppm. This abundance of silicon in the form of oxide can induce the idea that it is readily available, which would be a misleading idea since the production of this element is highly energy-consuming.

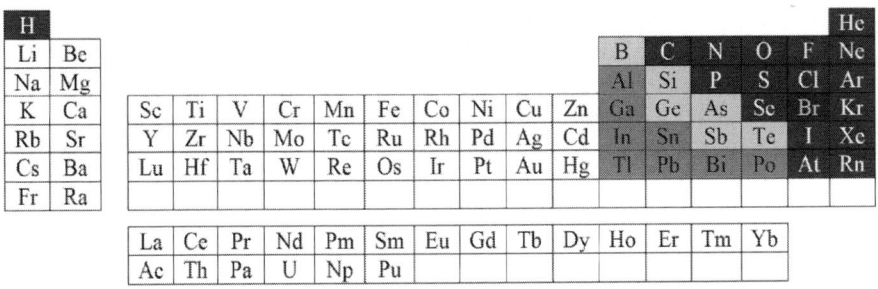

Figure 1.7. *Periodic table indicating metals, poor metals, metalloids and non-metals. For a color version of this figure, see www.iste.co.uk/valls/inorganic.zip*

From the perspective of metals, the history of humanity started with the bronze age (first furnaces), which was followed by the iron age (high temperature furnaces) and a further continuation could be the aluminum age (electrolysis of molten salts), but it may not be flattering to reduce the modern man to a period that would most certainly be chased away by another one, given that history tends to define our age as an achievement that is difficult to place in a finite continuum.

NOTE 1.2.– In our opinion, iron is an exceptional element. On the one hand, it is the main element of our planet; it contributes to the magnetic field that protects us and its most annoying property renders it respectable. Unlike aluminum, zinc, lead and many other elements that resist corrosion, iron always gets oxidized in the end and inevitably returns to its natural state (rust).

This obstinacy in getting oxidized, coupled with the relative ease in obtaining it, renders iron able to take part in the great cycles of our planet, as it is a known fact that if it can be reduced to object production, it will sooner or later return to the state of oxide as soon as man ceases to oppose to its oxidation.

The structuring of the classification (see Figure 1.8) allows the visualization of elements from the first (H) to the latest proposed element (plutonium for natural elements and oganesson for synthetic elements), when browsed from left to right, but it also indicates the order in which atomic subshells are filled.

This classification can be accessed on the site of the International Union of Pure and Applied Chemistry (IUPAC), which also indicates the values of atomic weights of the elements (*http://www.chem.qmul.ac.uk/iupac/AtWt/table.html*).

Rows of the periodic table are called periods as they define the period of variation of the properties of the elements. It is a specific type of periodicity, since it is 2, 8, 8, 18, 18 and 32 corresponding to the number of elements in each period.

	1	2	3	4	5	6	7	8	9	10	11	12	13	14	15	16	17	18
1	H																	He
2	Li	Be											B	C	N	O	F	Ne
3	Na	Mg											Al	Si	P	S	Cl	Ar
4	K	Ca	Sc	Ti	V	Cr	Mn	Fe	Co	Ni	Cu	Zn	Ga	Ge	As	Se	Br	Kr
5	Rb	Sr	Y	Zr	Nb	Mo	Tc	Ru	Rh	Pd	Ag	Cd	In	Sn	Sb	Te	I	Xe
6	Cs	Ba	Lu	Hf	Ta	W	Re	Os	Ir	Pt	Au	Hg	Tl	Pb	Bi	Po	At	Rn
7	Fr	Ra	Lr	Rf	Db	Sg	Bh	Hs	Mt	Ds	Rg	Cn	Nh	Fl	Mc	Lv	Ts	Og

| | | La | Ce | Pr | Nd | Pm | Sm | Eu | Gd | Tb | Dy | Ho | Er | Tm | Yb | | | |
| | | Ac | Th | Pa | U | Np | Pu | Am | Cm | Bk | Cf | Es | Fm | Md | No | | | |

Figure 1.8. *Current periodic table of elements*

NOTE 1.3.– The identification of elements 113, 115, 117 and 118 was only confirmed by IUPAC in 2015 and their final names were proposed in 2016. They are respectively nihonium, moscovium, tennessine (derived from Japan, Moscow and Tennessee, respectively) and oganesson (which renders homage to Y. Oganessian, a pioneer in the research on transactinides). The history of sciences is embedded in the periodic table through the names of elements, as it is easy to guess the origin of names such as californium, europium, indium, polonium, scandium or francium, but also of einsteinium, nobelium, curium or mendelevium.

It may be thought that a limit has been reached, but the quest continues, and it is hoped that after this last series of short-lived elements, significantly heavier elements will be created. Some of these super-heavy elements may be found in an island of hypothetical stability, in which they would exhibit a significantly higher half-life compared to neighboring isotopes ... time will tell!

The notion of period is also useful because the last period is the location of valence electrons, which explain the properties of the respective element. It should be noted that this does not mean that all electrons in the period are valence electrons, and zinc offers a perfect illustration in this sense: while its period contains two 4s-electrons and 10 3d-electrons, only the 4s-electrons are available and participate in the chemistry of this element, yielding only the Zn^{2+} ion.

The columns in the table are qualified irrespective of the chemical families, series or groups, but a certain number of precautions should be considered, for example for the first column, which comprises alkali metals (from lithium to francium) but does not include hydrogen, which is a non-metal. On the other hand, column 18 comprises noble gases (from neon to radon) and also helium, which is part of the s-block (column two) but its valence shell being complete, it is a noble gas, though this fact is often forgotten.

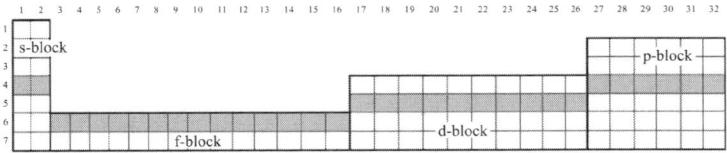

Figure 1.9. *Current extended classification (32 columns)*

Blocks can be visualized as a sequence of s, f, d and p (see Figure 1.9) if the classification is viewed in its non-truncated structure where the f-block is unfolded, corresponding to the four types of existing atomic orbitals. A block group denotes those elements whose same subshell is being completed or already complete. The only exception to this division into blocks is helium, which is above the column of noble gases, out of the s-block.

The most explicit rendering of the classification counts seven rows and 32 columns, which is a difficult to handle configuration. This explains the positioning of the f-block, which reduces its width by half and makes it easier to use.

The empty central part is always a problem for the chemist, who, similarly to nature, abhors vacuum. The following are therefore worth remembering:

– level 1 (or period 1) has only one type of subshell (1s);

– level 2 (or period 2) has two types of subshells (2s and 2p);

– level 3 (which is not period 3) has three types of subshells (3s, 3p and 3d), 3d being in period 4 since 4s-subshell is completed before 3d-subshell;

– level 4 counts four types of subshells (4s, 4p, 4d and 4f) distributed over the periods 4 (4s and 4p), 5 (4d) and 6 (4f);

– level 5 only counts the 5s, 5p, 5d and 5f subshells (the 5g-subshell has not been reached by natural elements) which are distributed over three periods, similarly to level 4;

– level 6 only counts 6s and 6p-subshells and level 7 only the 7s-subshell (and 7p-subshell with synthetic elements).

Therefore the empty spaces at the center of the classification result from the order of filling/completeness associated with its organization into columns and blocks.

1.2.3. *Analysis of various classifications*

1.2.3.1. *Classification according to Mendeleev*

When the first proposals of classification are analyzed, the uncertainties related to certain elements should be taken into account.

Taking up what is accurately presented in Figure 1.10 against a colored background, and reversing rows and columns and correcting the symbols, leads to 48 elements that are correctly positioned, out of which three were forecast (see Figure 1.11).

Except for several elements whose positioning is incorrect and for those that were not defined, this table has nevertheless the merit of presenting, though not explicitly stating, the periods and columns of the future classification.

The law of octaves has not yet been stated, but 7 years later (see Figure 1.12), Mendeleev would adjust his predictions and propose a format that highlighted this rule of octaves, which was a step toward the modern classification.

I	II	III	IV	V	VI
			Ti	Zr	P (Hf)
			V	Nb	Ta
			Cr	Mo	W
			Mn	Rh	Pt
			Fe	Ru	Ir
			Ni = Co	Pd	Os
H			Cu	Ag	Hg
	Be	Mg	Zn	Cd	
	B	Al	68 (Ga 69.7)	Ur (In)	Au
	C	Si	70 (Ge 72.6)	Sb	
	N	P	As	Sn	Bi
	O	S	Se	Te	
	F	Cl	Br	I	
Li	Na	K	Rb	Cs	Tl
		Ca	Sr	Ba	Pb
		45 (Sc)			

Figure 1.10. *Classification proposed by Mendeleev in 1869*

The d- and f-blocks still present a problem but the s- and p-blocks are correctly defined, and there is improved accuracy in the definition of the periodicity of properties, though the periods of the classification are not yet fully defined.

16   Inorganic Chemistry

H																	
Li	Be											B	C	N	O	F	
Na	Mg											Al	Si	P	S	Cl	
K	Ca	45	Ti	V	Cr	Mn	Fe			Cu	Zn	68	70	As	Se	Br	
Rb	Sr		Zr	Nb	Mo		Ru		Pd	Ag	Cd	In			Te	I	
Cs	Ba		Hf	Ta	W								Bi				

Figure 1.11. *Mendeleev's classification in the current classification grid*

Original presentations of the classification have employed various formats: spirals, squares, pyramids, etc., but these are poorly adapted to rapid transcription and effective sharing.

They are outshined by the current classification, whose predominance is easy to understand, as the following description will show.

	I	II	III	IV	V	VI	VII	VIII
1	H							
2	Li	Be	B	C	N	O		
3	Na	Mg	Al	Si	P	S	F	
4	K	Ca	44 (Sc)	Ti	V	Cr	Cl	
5	Cu	Zn	Ga	72 (Ge)	As	Se	Mn	Fe Co Ni Cu
6	Rb	Sr	Yt (Y)	Zr	Nb	Mo	Br	
7	Ag	Cd	In	Sn	Sb	Te	-	Ru Rh Pt Ag
8	Cs	Ba	Di, La	Ce			I	
9								
10			Er	La, Di	Ta	W		Os Ir Pt Au
11	Au	Hg	Tl	Pb	Bi			
12				Th		U		

Figure 1.12. *Periodic table proposed by Mendeleev in 1876*

1.2.3.2. *Construction of the current classification*

Let us build the classification of 94 natural elements based on their atomic numbers (not the atomic weights, obviously!) by lining up the chemically similar elements (alkaline elements and rare gases suffice).

Should the elements be arranged in one string and ranked by ascending atomic number, from 1 (hydrogen) to 94 (plutonium), and then cut the string after the noble gases, seven colored strips will be obtained (see Figure 1.13).

For good readability of Figure 1.13, the dotted line after hafnium and before astatine indicates the list of tantalum, tungsten, rhenium, osmium, iridium, platinum, gold, mercury, thallium, lead, bismuth and polonium.

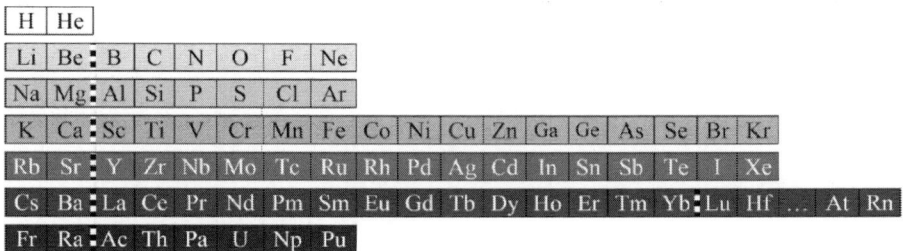

Figure 1.13. *Classification of elements in ascending order of atomic number*

Alkaline elements are lined up under hydrogen, and strips 2 and 3 as well as 6 and 7 are cut in order to align the noble gases and place the lanthanides and actinides under the table, to reduce its size (see Figure 1.14).

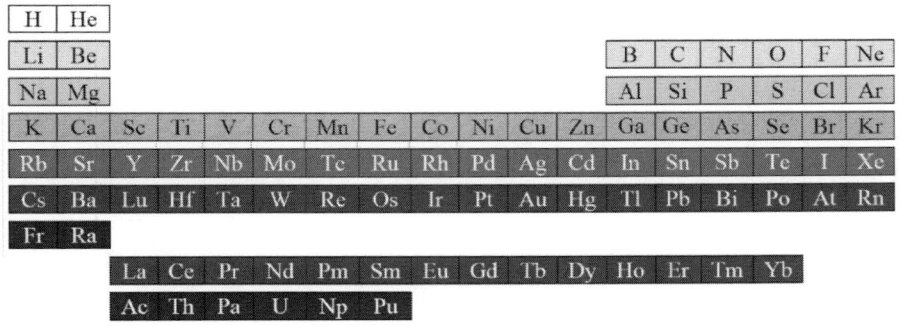

Figure 1.14. *Positioning of elements with reference to alkaline elements and noble gases*

Finally, helium is placed in the column of noble gases and the s- and d-blocks are separated to signal that f-block goes under the classification; the current structuring of the classification is thus obtained (see Figure 1.15) to which the synthetic elements recognized by the *International Union of Pure and Applied Chemistry* can be added, including the latest (nihonium, moscovium and tenessine) added in 2016.

This classification highlights synthetic elements in italics, non-metals on a light blue background and metalloids on a dark blue background.

It is possible to visualize the seven periods vertically numbered from 1 to 7 and the 18 columns horizontally numbered from 1 to 18. The use of atomic number and the alignment of elements with similar properties facilitate the highlighting of blocks and subshells and subsequently the definition of periods, but this distribution is even more resourceful than it may seem.

It is possible to visualize the electron subshell filling order (with a few exceptions), the four blocks (s, f, d and p, in this order) that have been framed, the periods (horizontally) from 1 to 7, the families of elements (vertically) and the nature of elements (metals, metalloids and non-metals) which evidences the diagonalization of the classification, a phenomenon that will be subsequently detailed (see section 1.3.2).

Though not present, the roman numerals introduced by Moseley can be assigned, for example, to the ions formed by various columns. Indeed, the monatomic cations refer to columns from 1 to 12, as well as to poor metals (of p-block), several metalloids (arsenic, antimony and tellurium) and the f-block. Monatomic anions only relate to the p-block, as monatomic metals yield exclusively cations.

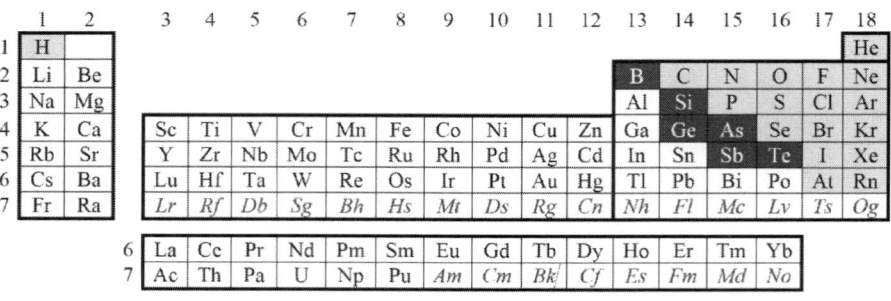

Figure 1.15. *Current classification highlighting metals, non-metals and metalloids. For a color version of this figure, see www.iste.co.uk/valls/inorganic.zip*

The qualitative definition of all natural elements is now available, but it offers no quantitative information.

If interested in the distribution of elements on Earth, one can consult studies on their abundance, but this approach would lead from the certainties of the classification to the uncertainties of modern science, with its sometimes contradictory mix of citations and theories.

The study of the distribution of elements indeed started as late as 1920, when Goldschmidt [GOL 26] formulated the first hypotheses, and it is still ongoing.

1.2.4. Abundance of elements

The universe is mainly formed of hydrogen (90%), helium (9%) and several other elements with a significant presence: oxygen, carbon and neon [KRA 67]. When stars and planets form, the diversity of atoms in these celestial bodies increases, and selective distribution of these elements in various layers can be observed.

Understanding this selective distribution of elements involves knowledge of geochemistry, which will be succinctly approached here.

1.2.4.1. Abundance of elements in the Earth's crust

While certain elements such as gold, silver or copper (and also platinum, iron-nickel in meteorites, tin and mercury) are native elements and can be found in nature in metal form, most of the elements are stable oxides that are difficult to reduce (e.g. alkaline elements or aluminum); therefore research on elements has not been a simple task.

Moreover, the element with the strongest presence in the Earth's crust is not a metal, but oxygen (through all the oxides), followed by silicon (oxidized in silicates) while the first metal, aluminum (in the form of oxide), ranks only third.

In fact, SIAL is the name (a combination of the first letters of silicon and aluminum) assigned to the Earth's crust, which is mainly formed of aluminosilicates and aluminum silicates.

Metals make up the majority of the classification (68 elements) but their quantity in the Earth's crust is not proportional with their number, quite the opposite being true (see Figure 1.16). The vast majority of rocks and minerals (98%) are composed of eight elements, (listed in descending order: oxygen, silicon, aluminum, calcium, iron, sodium, magnesium and potassium) six of which are metals, while all the other elements are dispersed or rare [HAY 15].

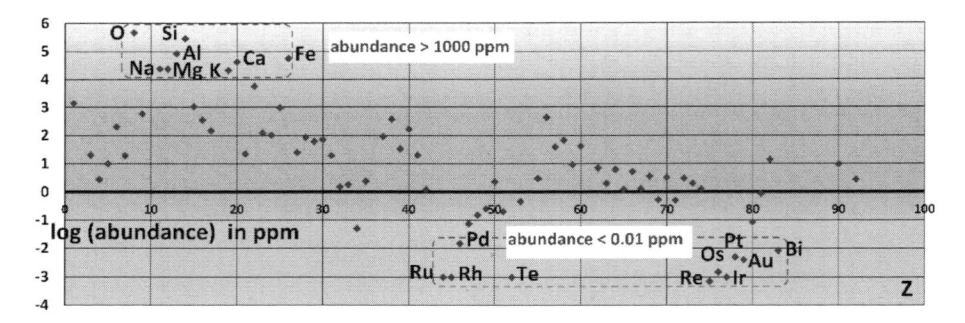

Figure 1.16. *Abundance of elements in the Earth's crust depending on their atomic number. For a color version of this figure, see www.iste.co.uk/valls/inorganic.zip*

The Earth's crust is the most accessible to human activity; it is a place of continuous, though on the geological scale, mixing of elements that date back to the beginnings of our planet. It is therefore natural to turn to geochemistry when in search for answers. This discipline studies the cycles through which most of the chemical elements are alternatively brought to the Earth's surface and back in its depths, and applies the tools and concepts of chemistry to the study of the earth and its sediments.

In the 1920s, Vladimir Vernadsky and Victor Goldschmidt [GOL 26] proposed a first vision of geochemistry and classified the elements into four categories (see Figure 1.17):

– lithophiles, which exhibit an affinity for oxygen, are very present in the Earth's crust mainly in the form of aluminates and silicates. They include oxygen, silicon, aluminum, calcium, sodium, magnesium and potassium (iron, which is simultaneously lithophile and siderophile, is also present at the surface and is ranked 4th);

– atmophiles (hydrogen, carbon, nitrogen and noble gases) which are mainly present in the atmosphere in the form of dinitrogen, dioxygen, carbon dioxide, noble gases and water (in the oceans), have predominant affinity for fluid phases;

– siderophiles, which have affinity for iron and have become scarce on the surface as result of their migration to deeper layers, close to the Earth's core. Platinum, gold, rhenium, osmium, ruthenium, rhodium, palladium and above all iridium (present at 1 µg per kg) are therefore the rarest metals in the Earth's crust. Manganese, which is simultaneously siderophile and lithophile, is the only one having a significant presence, with almost 1 mg per kg, but this is far from aluminum, which is over 80 times more abundant;

H																	He
Li	Be		atmophiles		lithophiles							B	C	N	O	F	Ne
Na	Mg		chalcophiles		siderophiles							Al	Si	P	S	Cl	Ar
K	Ca	Sc	Ti	V	Cr	Mn	Fe	Co	Ni	Cu	Zn	Ga	Ge	As	Se	Br	Kr
Rb	Sr	Y	Zr	Nb	Mo	Tc	Ru	Rh	Pd	Ag	Cd	In	Sn	Sb	Te	I	Xe
Cs	Ba	Lu	Hf	Ta	W	Re	Os	Ir	Pt	Au	Hg	Tl	Pb	Bi	Po	At	Rn
Fr	Ra																

La	Ce	Pr	Nd	Pm	Sm	Eu	Gd	Tb	Dy	Ho	Er	Tm	Yb
Ac	Th	Pa	U	Np	Pu								

Figure 1.17. *Nature of the elements present in the Earth's outer crust according to Goldschmidt [GOL 26] (underlined elements belong to several categories). For a color version of this figure, see www.iste.co.uk/valls/inorganic.zip*

– chalcophiles are frequently encountered in association with sulfur due to a higher affinity with this element than for oxygen. Their presence on the surface is reduced because sulfides are denser than silicates.

This classification is not standardized, but it has the merit of being the first one proposed, and it frequently happens that when an element is part of two families, its behavior places it sometimes in one family, and at other times in the other:

– iron is paradoxically at the same time chalcophile, lithophile and siderophile, which accounts for its abundance in all of Earth's layers;

– carbon is generally considered an atmophile, as its solid forms (coal and carbonates) originate in atmospheric carbon dioxide, but it is siderophile in the absence of oxygen;

– phosphorus is generally considered lithophile, but it is also siderophile in the absence of oxygen.

22   Inorganic Chemistry

It can be noted that the Earth's crust is rich in lithophile elements (which form low density solid oxides that have been concentrated in the outer layers of the Earth) but it is strongly lacking in siderophile elements (which are high-density elements and are carried along with the iron to the core) and in atmophile elements (which are too volatile to be integrated into the Earth's mass). It is also scarce in chalcophile elements, which are denser than the oxides formed by lithophiles.

This first classification has the merit of precedence and comprises all the elements whose distribution can be analyzed based on other groups, such as oxides, sulfides, sulfates, carbonates, silicates, halides and hydroxides.

Depending on the inventory requirements, an analysis of magmatic rocks (basalt, granite, gabbro, etc.), sedimentary rocks (limestone, dolomite, sandstone, etc.), metamorphic rocks (schist, marble, gneiss, etc.) can be drawn (see Figure 1.18).

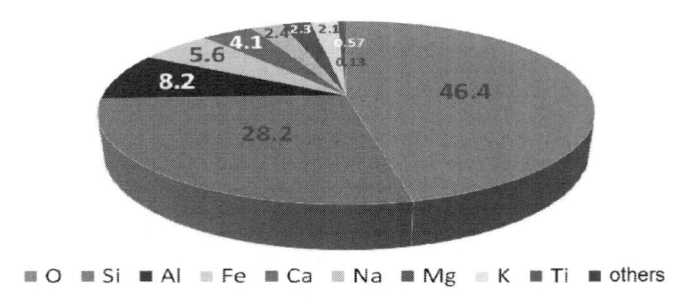

■ O ■ Si ■ Al ■ Fe ■ Ca ■ Na ■ Mg   K ■ Ti ■ others

Figure 1.18. *Composition (%) of the Earth's crust. For a color version of this figure, see www.iste.co.uk/valls/inorganic.zip*

Non-silicates are classified depending on their chemical composition: halides, the most common of which in nature is sodium chloride, NaCl; sulfides and sulfosalts, which very often constitute ores, containing many metals (pyrite, FeS_2; chalcopyrite, $CuFeS_2$; galena, PbS; sphalerite, ZnS); carbonates, which are mainly represented by calcite ($CaCO_3$); sulfates, with three common species, gypsum ($CaSO_4$, $2H_2O$), anhydrite ($CaSO_4$) and barytine ($BaSO_4$); phosphates, only one species of which is common, apatite $[Ca_5(PO_4)_3(OH, F$ and $Cl)]$; and finally, oxides and hydroxides, with the group of spinels, ilmenite ($FeTiO_3$), rutile (TiO_2), etc.

The mantle contains essentially silicon and magnesium, and it is also called SIMA, a term resulting from the first letters of silicon and magnesium.

The Earth's core is liquid and contains iron and sulfur, similar to the grain which also contains nickel. The grain is thought to be solid, because its temperature is about 3000 °C, its pressure reaches 4 Mbar and its depth is over 5000 km (see Table 1.2).

Shell	Depth (km)	Density (g/cm^3)	Chemical elements
Continental crust	0 – 35	2.7 – 3.0	Si and Al
Oceanic crust	0 – 10	2.9 – 3.2	Si, Al and Mg
Upper mantle	35/10 – 670	3.4 – 4.4	Si, Mg and Ca
Lower mantle	670 – 2890	4.4 – 5.6	Si, Mg, Fe and Ca
Outer core	2890 – 5100	9.9 – 12.2	Fe, Ni and S (liquid state)
Inner core	5100 – 6378	12.8 – 13.1	Fe, Ni and S (solid state)

Table 1.2. *Nature of the elements present in the terrestrial globe [MIC 16]*

1.2.4.2. *Abundance in the human body*

After this account of the elements contained by the Earth and their distribution, and after the attempt to understand the mechanism of this distribution, it is natural to describe the elements chosen by organic life for its emergence and development on Earth.

Six elements account for 98.5% of the human body mass (oxygen 65%, carbon 18%, hydrogen 10%, nitrogen 3%, calcium 1.5% and phosphorus 1%) and are above 1% and then seven elements constitute a minority below 0.5% (potassium 0.4%, sulfur 0.3%, sodium 0.2%, chlorine 0.2%, magnesium 0.1%, iodine 0.1% and iron 0.1%). Several elements play an important enzymatic role (lithium, strontium, aluminum, silicon, lead, vanadium, arsenic and bromine), but they account for less than 0.1% of the body mass. As for the living world as a whole, we can suppose that it involves 27 elements, which is only a tiny part of the 94 existing elements.

Life has therefore chosen mainly non-metals that are far more versatile in terms of the nature (covalent or ionic) and type (simple or multiple) of bonds they form. Due to the water in the human body, oxygen remains the pillar of elements. It is followed by carbon, nitrogen and phosphorus, which are likely to simultaneously account for most bonds, i.e. four and three. While calcium plays an unquestionable role, it is due to its use in the structure of the human skeletal system that it ranks first among metals.

Human life depends on high levels of consumption of oxygen (drinking and respiration), carbon, hydrogen, nitrogen, calcium and phosphorus (food), but various lifestyles may entail far higher demands. For example, French people have an average annual consumption per person of 250 kg of iron, 13 kg of aluminum, 8 kg of copper, 5 kg of zinc, 3.6 kg of lead and 1 kg of nickel [BLA 11].

Pollution, hunger and many other imbalances would be reduced by adopting lifestyles that are more respectful of nature, but quite distant from the currently prevailing consumerist principles.

For example, it is worth noting that during the last 30 years there has been a significant increase in the consumption of metals, both in terms of quantity and number.

The development of new technologies has led to an increase in the demand of *high tech* metals, a trend that is currently applicable to all metals, even to the less stable and rarest ones, such as polonium and astatine, which are not found in nature and are being produced by nuclear reactors [BIH 10].

1.3. Reading the classification

In its structured form, the classification can be used as a source of information, since there is certain continuity in the evolution of the properties of elements. For example, values of many quantities can be observed to vary periodically, by reading the classification; this is not a simple task, as many phenomena need to be simultaneously taken into account.

There are therefore keys to reading and a definition of the most helpful in understanding crystals will be provided here.

Four quantities will be analyzed, each of which evidences either the evolution or the existence of a property:

– atomic radius (Ra), which unambiguously shows periodicity;

– electronegativity (χ), which can be associated with the diagonalization of classification;

– ionization potential (IP_n or I_n), which evidences the stability conferred to the electronic structures by fully occupied, empty or semi-occupied subshells;

– electron binding energy (BE), which confirms the stability conferred by fully occupied, empty or semi-occupied subshells.

Evolutions, tendencies or aptitudes can be proposed, rather than precise equations of curves, and this will be useful in explaining and analyzing most of the properties observed.

1.3.1. *Atomic radius*

It is difficult to define the atomic radius, as it corresponds to the distance from the nucleus at which there is the maximum probability of finding valence electrons (predicted by Bohr's model for hydrogen). Several calculation methods are used [SLA 64]. It is indeed impossible to experimentally determine the radius of an isolated atom (r_S).

Different types of calculation yield various values for the same atom (covalent radius (r_C) [SUT 65], metallic radius (r_M) [PAU 60], van der Waals radius (r_{vdW}) [BON 64]), which are also provided in this book. For consistency reasons the values proposed in the Handbook (r_H) [HAY 15] will be used (see Table 1.3).

When an atom bonds with another atom, the electron cloud arrangement is strongly modified by the bond. Depending on the nature of the latter, a covalent, ionic or metallic radius is defined, which facilitates the explanation or prediction of the evolution of interatomic distances and of the bond energy for the respective elements.

Element	r_S [SLA 64]	r_C [SUT 65]	r_M [PAU 60]	R_{vdW} [BON 64]	R_H [HAN 15]
Li	145	134	122.5	180	130
Na	180	154	157.2	230	160
Mg	150	145	136.4	170	140
Si	110	118	117.3	210	114
P	100	110	110	185	109
S	100	102	104	180	104
Zn	135	120	124.9	140	120

Table 1.3. *Atomic, covalent, metallic, van der Waals radii and values of covalent radii of several atoms, according to the Handbook [HAY 15]*

Figure 1.19 offers a representation of the covalent radii of the first 70 elements and gives us an idea of the evolution of this property in the classification. The blocks are assigned the following symbols: ▲ for s, ● for p, ◆ for d and O for f.

The size of the atoms depends on the number of protons in the nucleus (Z), which attract electrons and therefore lead to a decrease in atom size. An increase in Z entails an increase in the number of electrons; therefore, their repulsive interactions and thus the size of the atom increase. These two opposing phenomena generate a periodic variation of quantities along the rows of the classification. It is worth keeping in mind that this period is not constant; it is in fact a pseudo-period, which has for a long time been a source of difficulties that were solved by structuring the classification (see section 1.2.2).

These are two opposing phenomena and nucleus–electron attraction generally prevails over electron–electron repulsion, except for the change of period. Moreover, atomic radii only vary by a factor of 10 while Z varies by a factor of 100.

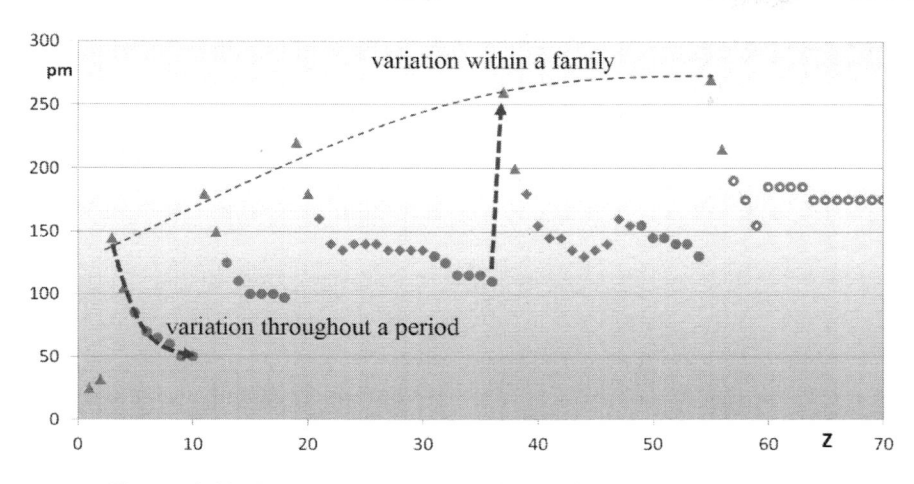

Figure 1.19. *Atomic radii calculated for the first 70 elements. For a color version of this figure, see www.iste.co.uk/valls/inorganic.zip*

When the period changes a new shell forms and there is strong increase in size while the attraction of the newly added proton is not sufficient to compensate for this increase in atom size. For example, the atomic radius of neon (10th element) is 62 pm and that of sodium (11th element) is 160 pm. Furthermore, there is no creation of a new shell, but filling of the subshells of the respective level, and the increasing attraction of the nucleus leads to the compression of the electron cloud of atoms up to noble gas (from 140 pm for magnesium to 101 pm for argon).

The evolution of the atomic radius in the classification can be summed up by noting that it decreases throughout a period, increases throughout a column and strongly increases upon change of period (indicated by the values for periods from 4 to 5). This description is symbolically represented by the three arrows in Figure 1.20.

The observation of the evolution of atomic radii highlights the alkaline elements, which are the largest atoms of each period, and the noble gases, which are the smallest, as well as a diagonal of lithium, magnesium, gallium, tin and bismuth, for which the value of the atomic radius is close to 140 pm (blue line).

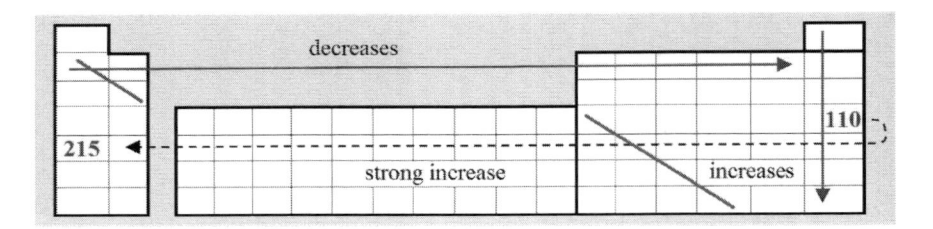

Figure 1.20. *Current classification*

NOTE 1.4.– An opposition of the effects described and therefore a weak variation of atomic radius can be expected along the diagonals descending to the right. Indeed, the diagonal starting from lithium (130 pm) and going through magnesium (140 pm), gallium (123 pm), tin (140 pm) and bismuth (150 pm) yields an average value of 137 ± 13 pm, which amounts to 10% variation. Though not confirmed, the existence of diagonalization of the classification can be inferred (it will only be confirmed based on values of electronegativity).

The diagonal that links boron to astatine does not evidence diagonalization, and the following values of atomic radii are observed: 84, 114, 120, 137 and 148 pm, when they should be more or less constant. Likewise, values 75, 109, 118, 136 and 146 can be observed from carbon to radon. It is therefore all the more difficult to conclude that 35 atomic radii are comprised between 123 and 150 pm over 94 elements.

1.3.2. *Electronegativity*

Electronegativity (χ) is the tendency of an atom engaged in a bond to attract electrons participating in that bond. Electronegativity was initially defined by Pauling, who used it to characterize the chemical bond [PAU 60]. It only makes sense for bonded atoms; therefore it is not a property of noble gases (except for krypton and xenon, for which compounds have been obtained).

There is no one definition of electronegativity and various accepted definitions yield several electronegativity scales. The three most well-known electronegativity scales are those of Mulliken, Allred and Rochow, and Pauling, the latter being the one more usually employed by chemists.

While electronegativity characterizes the aptitude of atoms to gain or lose electrons, it is still an abstract notion that does not directly rely on the measurement of an experimental quantity. It is an important chemical property, as it is involved in explaining the formation of ions and molecules.

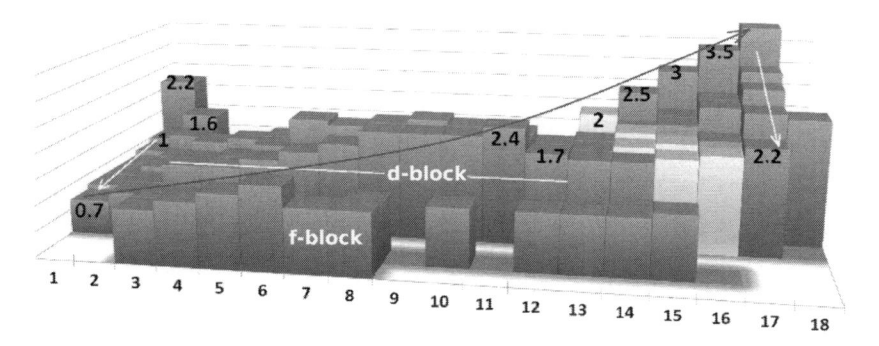

Figure 1.21. *Electronegativity in the classification*

Figure 1.21 summarizes the values of electronegativity for elements in the periodic table according to the Pauling scale. The periodicity described for the atomic radius is readily verified but the evolution of electronegativity in the classification provides further information and shows that:

– as a general rule, electronegativity decreases in a column when Z increases (note the arrows on the alkaline or halogen elements);

– in a period (row in the classification) electronegativity generally increases when Z increases (the cross-cutting arrow cumulates the two variations);

– on the right side there are electronegative non-metals, which attract electrons when they form a bond (they are oxidizers); hydrogen, on the left, should also be mentioned, though it is a non-metal;

– electropositive metals (reducing elements) can be observed on the left; they always yield cations if they are monatomic (to be more precise, they never yield monatomic anions).

An irregular stairway can be imagined, whose steps go in both directions and climb continuously from cesium to fluorine whatever the step taken, several steps being slightly shifted downwards (for example, those of zinc and cadmium, column 12).

The values of electronegativity can be readily memorized by smoothing out values and by considering the following observations:

– metalloids have an intermediate electronegativity that can be associated with a value of 2 (indicated in white) as their name indicates; they make the transition between metals and non-metals and show the diagonalization of the classification. It is according to this type of diagonal that different groups proposed are separated relative to electronegativity (see Figure 1.22);

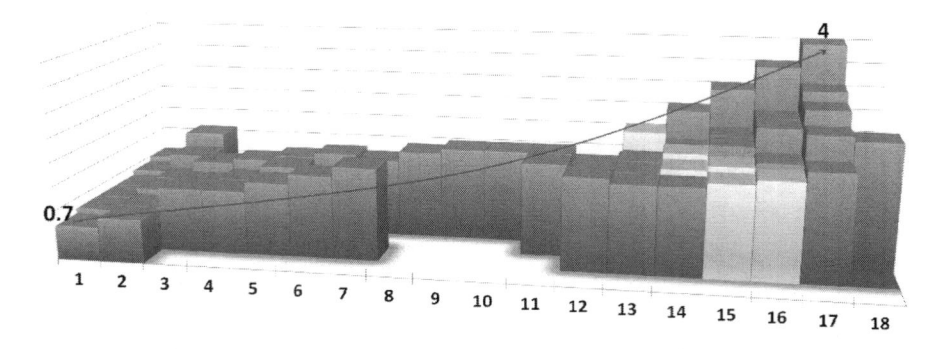

Figure 1.22. *Evolution of electronegativity in the periodic table*

– non-metals to the right of metalloids have an electronegativity above 2;

– metals to their left generally have an electronegativity below 2 (except for several metals, which never yield monatomic anions);

– extremes F (4), Fr (0.7), H (2.2) and Au (2.4) are considered and 1.8 is recorded as an average value for the common metals;

– the least electronegative metals are qualified as electropositive, their electronegativity being below or equal to 1;

– metals whose average electronegativity is 1.4 are qualified as metal medians;

– finally, the f-block, which is located between the two last categories, has an intermediate average value of 1.1 for the 4f-subshell and 1.3 for the 5f-subshell.

Electronegativity highlights the metalloids and the diagonalization phenomenon and it also stresses the extremes such as fluorine, francium or gold and the atypical elements such as hydrogen (see Figure 1.23).

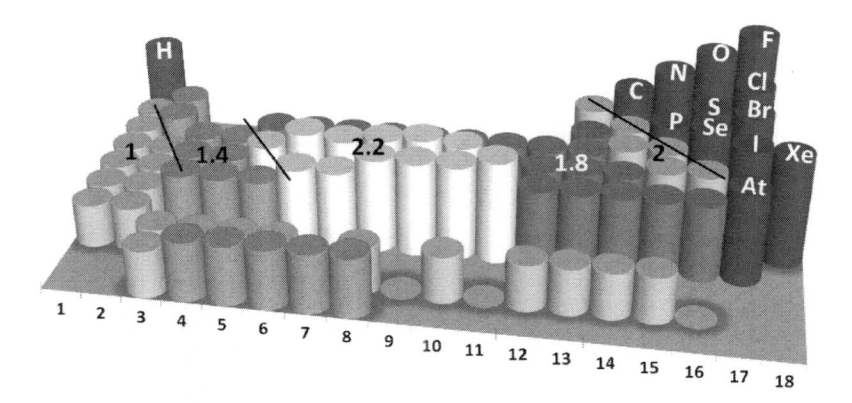

Figure 1.23. *Smoothed evolution of electronegativity in the classification. For a color version of this figure, see www.iste.co.uk/valls/inorganic.zip*

NOTE 1.5.– It is worth noting that the electronegativity of the second period evolves from 1 to 4 in steps of 0.5 (except for beryllium, which has a value of 1.55 instead of 1.5). It is therefore easy to retain the values of electronegativity by memorizing different variations as applicable to the rows and columns of the classification.

1.3.3. *Ionization potential*

The ionization potential (IP_n or I_n) is the energy required to force out of an atom of gas one or several electrons.

IP_1 corresponds to the first ionization potential: $A_{(g)} \rightarrow A^+_{(g)} + 1\ e^-$

A second ionization potential IP_2 is involved if a second electron is removed and so on.

Orbital energy is equal to the absolute value of the corresponding ionization energy, a result which is also known as Koopmans' theorem [KOO 34].

A very satisfactory approximation of an atom's first ionization energy can be obtained directly from the energy of the last occupied orbital. Conversely, knowing the first ionization energy makes it possible to position the last occupied orbital on an energy diagram.

Figure 1.24 shows that IP_1 increases over a period (from the left to the right side of the classification) and decreases when going downwards along a family (from top to bottom of the classification).

The larger the atom, the farther away from the nucleus are the outermost electrons and the weaker the attraction forces exerted on them, which explains why alkaline elements are easy to ionize. On the other hand, the ionization of noble gases and of halogens, up to alkaline elements, is energy consuming.

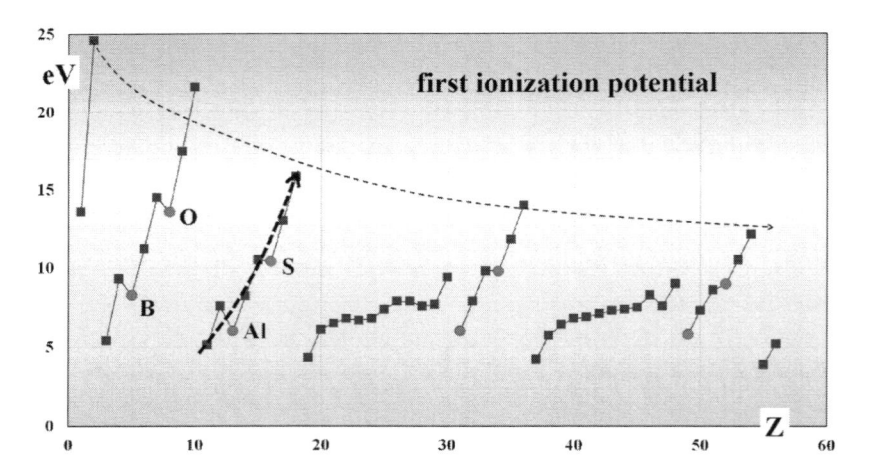

Figure 1.24. *Evolution of the first ionization potential in the classification. For a color version of this figure, see www.iste.co.uk/valls/inorganic.zip*

This type of diagram evidences the higher stability of the configurations that have fully occupied subshells, and for example IP_1 (for boron, aluminum, etc.) is lower than expected because when the atom loses one electron, its np subshell is emptied, while its ns shell is occupied, which yields stability. Likewise, IP_1 (for oxygen, sulfur, etc.) is lower than expected because when the atom loses one electron, the np subshell is half occupied, which yields stability.

The elements concerned in this decrease in ionization potential are highlighted by a different color and a round mark (see Figure 1.24). This phenomenon can be noted for all the periods, but it rapidly fades out.

Figure 1.25 allows the visualization of two columns (13 and 16 in light blue) which present an anomaly related to the fully occupied s shell or the half-occupied p shell, respectively, when the elements lose one electron.

The value of the first ionization potential of hydrogen is abnormally high, as it has a half-filled valence shell (being therefore relatively stable). Its valence shell contains only one electron, which is difficult to remove, since this would entail removing all its electrons.

Moreover, its electronegativity is far more significant than that of neighboring atoms, as it is a non-metal which is close to carbon (in the next period), which has in turn a half-filled valence shell.

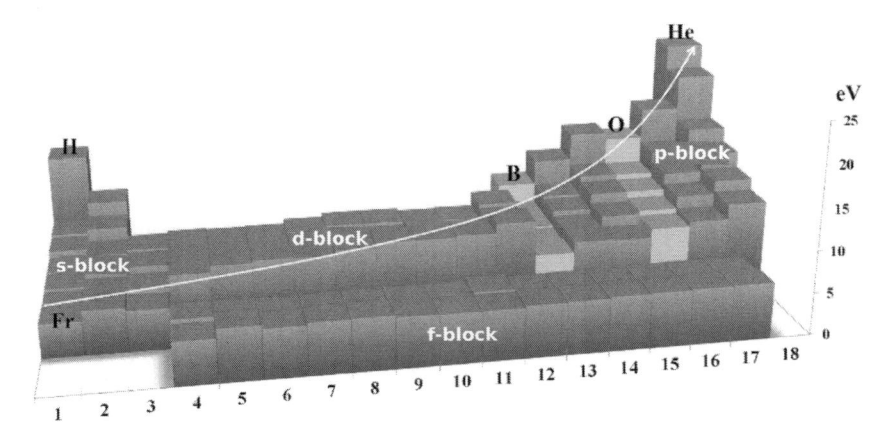

Figure 1.25. *Evolution of the first ionization potential of elements. For a color version of this figure, see www.iste.co.uk/valls/inorganic.zip*

The first ionization potential IP_1 is highest for noble gases, whose fully filled valence shell is exceptionally stable, and decreases to the lowest value for alkaline elements, which very easily lose one electron, and which yields a very stable M^+ cation, which is isoelectronic to the rare gas preceding it.

NOTE 1.6.– A very simple method can be devised for obtaining a ranking of elements based on their first ionization energy, by observing the classification. Similar to electronegativity, a stairway can be visualized, whose lowest and highest steps are those of francium and helium, respectively, taking into account hydrogen (a step higher than normal) and columns 13 and 16 (steps shifted downward). In order to range the elements in ascending order of their ionization potential, it suffices to climb the steps of this imaginary stairway!

The first ionization potential evidences periodicity, as well as columns 13 and 16, and it is the only parameter (together with the electron binding energy) to show the stability of fully or half-filled subshells. Hydrogen has a prominent position in this case too, similar to the other quantities, and it is helium, together with its noble gases family, that has the highest ionization potential.

1.3.4. *Electron binding energy*

Electron binding energy (BE) or electron affinity of an atom is the energy released when an element acquires one electron. Based on this definition, this quantity is generally negative, and the higher it is, the stronger the tendency of the atom to acquire one electron:

$$A_{(g)} + 1\ e^- \rightarrow A^-_{(g)} \qquad BE < 0 \text{ in general}$$

If the element acquires an electron and releases energy, the variation of enthalpy is negative. The values of electron binding energy are conventionally considered in absolute values, so that the higher the electron binding energy, the better the element is at acquiring electrons. A positive value presents no interest, as the electron is difficult to bind, and a zero or negative value (depending on tables) is therefore indicated.

Unlike the ionization energy, which is always positive, electron binding energy can be negative; but if it is positive, then the neutral atom is more stable than the anion.

When trying to understand the evolution of electron binding energy, it is recommended to keep in mind several phenomena, as it may seem anarchic at first observation (see Figure 1.26). When reading the classification from left to right one notes that it increases, having minima for the fully filled subshells (columns 2 and 18) or half-filled subshells (15). The maximum corresponds to the next to last column of a block as the electrons fill the last subshell of the element by accepting an electron and this yields a very stable anion.

When d-block is inserted (see Figure 1.27) reading becomes more complex, but in the light of the phenomena observed for s- and p-blocks, the following conclusions can readily be drawn:

– electron binding energy increases when going from left to right over p- and d-blocks and varies to a relatively little extent within one family;

– the last column in every block (2, 12 and 18) corresponds to a fully filled subshell and the value is null; the next to last column of every block (1, 11 and 17) corresponds to a maximum, as the elements have a complete subshell from having acquired an additional electron;

– the central columns (7 and 15, and not for the s-block, which only counts two) correspond to half-filled subshells and have a minimum value.

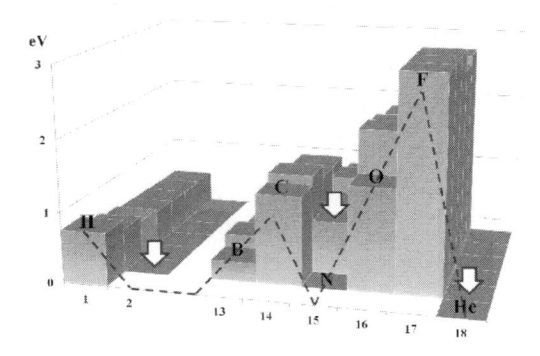

Figure 1.26. *Electron binding energy of the elements of s- and p-blocks*

36 Inorganic Chemistry

Unlike the previous quantities, electron binding energy highlights neither periodicity nor hydrogen, but it evidences columns 2, 12 and 18 on the one hand, and 7 and 15 on the other hand.

The four quantities described facilitate the association of certain columns or elements with a characteristic and therefore they can be memorized in relation with that particular characteristic. As already mentioned, there is an affective way of memorizing, the classification, and this knowledge of properties of elements helps in distinguishing them from the whole and therefore personalizing them. We have underlined the characteristics of eight columns, including those of metalloids, of highly electronegative metals and of many elements such as hydrogen, fluorine, cesium, gold, etc. and it is difficult to think of one property without immediately associating it with a series of elements.

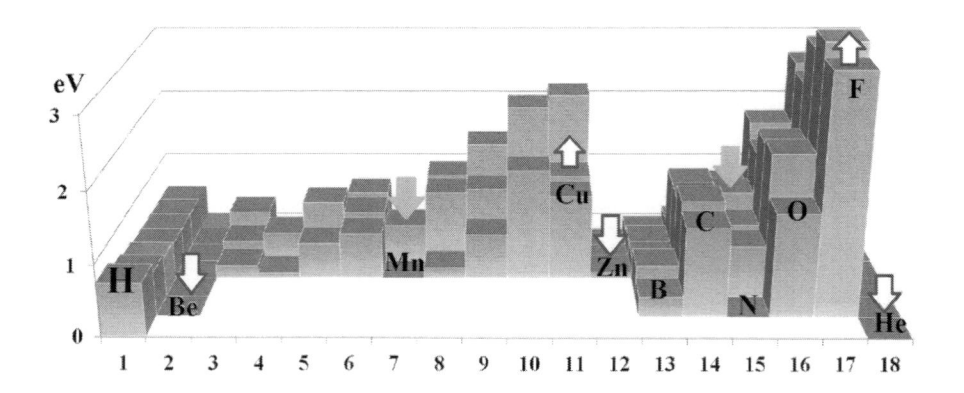

Figure 1.27. *Electron binding energy of the elements of s-, p- and d-blocks*

NOTE 1.7.– It should be noted that concluding that copper is stable by accepting one electron because the electron binding energy is relatively high, is erroneous reasoning.

In fact, first ionization potential involves values from 5 to 25 eV, while the values of the electron binding energy are ten times lower, ranging between 0.5 and 3 eV. The latter phenomenon occurs, but it is secondary in relation to the former. For example, it allows us to explain several exceptions to the Madelung rule or electron rearrangements in ions.

In the absence of knowledge of the above-described phenomena it is surprising to note that the addition of one electron to nitrogen leads to lower stability, while the opposite is true for carbon (as reflected by the value of its electron binding energy)!

1.4. Understanding ions through the classification

The description provided of some properties of the elements will be used to understand the nature of ions and their size and polarizability, which will be useful in understanding ionic crystals.

1.4.1. *The nature and valence of ions through the classification*

The classification can be summarized in seven horizontal lines, called periods, and 18 columns that can define the families (all the afore mentioned necessary precautions being considered). Families have in general similar chemical properties, as they have the same number of valence electrons and if they belong to the same group of elements, metals, non-metals and metalloids, they yield the same type of ions.

As Figure 1.28 readily shows, metals form only cations, while non-metals can form both anions and cations. It can be noted that the elements in columns 1 to 8 form cations with 1 to 8 charges, respectively, and hydrogen forms one anion and one cation with 1 charge.

The same is applicable to columns 11 to 17 (cations with 1 to 7 charges) with several exceptions related to the strong electronegativity of oxygen and fluorine (which only form anions).

Columns 15 to 17 form anions with, respectively, 1 to 5 charges only when the s-shell is concerned.

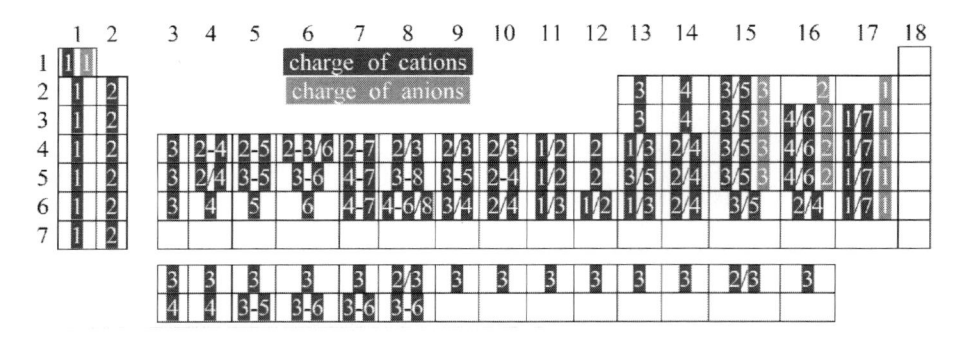

Figure 1.28. *Charge of ions corresponding to elements. For a color version of this figure, see www.iste.co.uk/valls/inorganic.zip*

Lanthanides form M^{3+} ions, and so do actinides, with the exception of actinium and thorium (only six are natural) and these are two families that read horizontally, although several elements form ions with 2 to 6 charges.

Generally speaking, all metals form monatomic cations (Na^+, Fe^{2+}, Al^{3+}, etc.) and some also form oxygenated anions when they exhibit a high oxidation level (CrO_4^{2-}, MoO_4^{2-}, MnO_4^-, etc.).

There are few monatomic anions in the classification, but the association between cations and oxygen atoms yields many highly stable negative polyatomic ions (CO_3^{2-}, MnO_4^-, NO_3^-, SO_4^{2-}, etc.).

Likewise, many elements of the d-block form M^{2+} ions (they empty the 2s orbital) or M^{3+} ions, but they are divided into families, as they also form cations with 1 to 8 charges (which are either real or virtual, depending on the oxidation states).

Depending on their charge, ions can be presented as follows:

– columns 1 (IA) and 11 (IB) form ions with 1 charge;

– columns 2 (IIA) and 12 (IIB) form ions with 2 charges, as well as many elements of the d-block, which lose their s-electrons;

– columns 3 (IIIA) and 13 (IIIB) form ions with 3 charges, as well as several elements of the d-block and many elements of the f-block;

– columns 15 (VA), 16 (VIB) and 17 (VIIB) form anions with, respectively, 3, 2 and 1 charges for non-metals;

– there are 76 elements that have ±1, ±2 and ±3 charges, so there are only several specific elements that form cations with 4, 5, 6, 7 and 8 charges;

– columns 4 (IVB) and 14 (IVA) form ions with 4 charges (for metals) and several d- and f-elements;

– columns 5 (VA) and 15 (VB) form ions with 5 charges;

– column 6 (VIA) forms ions with 6 charges;

– column 7 (VIIA) forms ions with 7 charges;

– column 8 forms ions with 8 charges (except for iron);

– columns 9 and 10 form cations with 3 to 5 charges;

– column 18 (VIIIB) does not form any ions.

Therefore, the position of an element in the classification indicates the type of ion that the element is likely to form.

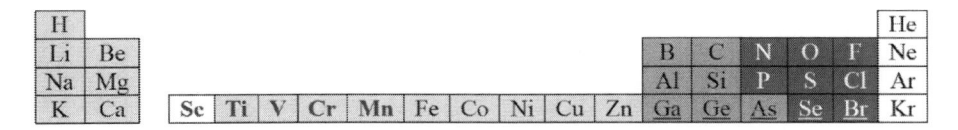

H																	He
Li	Be											B	C	N	O	F	Ne
Na	Mg											Al	Si	P	S	Cl	Ar
K	Ca	Sc	Ti	V	Cr	Mn	Fe	Co	Ni	Cu	Zn	Ga	Ge	As	Se	Br	Kr

Figure 1.29. *Behavior of elements in the periods from 1 to 4 during ion formation. For a color version of this figure, see www.iste.co.uk/valls/inorganic.zip*

The first four periods are represented in Figure 1.29, and the following can be noted:

– some atoms empty the s-subshell and form an ion, for example: H^+, Li^+, Be^{2+}, Na^+, Mg^{2+}, K^+, Ca^{2+}, Ti^{2+}, V^{2+}, Cr^{2+}, Mn^{2+}, Fe^{2+}, Co^{2+}, Ni^{2+}, Cu^{2+} and Zn^{2+} (marked in light blue);

– other atoms empty their s- and p-subshells, for example B^{+III}, C^{+IV}, Al^{3+} and Si^{+IV} (in medium blue), but some maintain a complete d-subshell, such as Ga^{3+}, Ge^{+IV} and As^{+IV} (underlined);

– some atoms empty their valence shell (marked in blue and boldfaced) such as $\mathbf{Sc^{3+}}$, $\mathbf{Ti^{4+}}$, $\mathbf{V^{+V}}$, $\mathbf{Cr^{+VI}}$ and $\mathbf{Mn^{+VII}}$, or $\mathbf{N^{+V}}$, $\mathbf{P^{+V}}$ and $\mathbf{S^{+VI}}$ (some only exist in the polyatomic form, such as MnO_4^-, CrO_4^{2-}, SF_6, N_2O_5, etc.);

– some atoms empty the p-subshell (underlined) such as $\underline{Ga}^+$, $\underline{Ge}^{2+}$, $\underline{As}^{+III}$, $\underline{Se}^{+IV}$ and $\underline{Br}^{+V}$;

– some atoms fill their p-subshell, these being the most electronegative, such as O^{2-}, F^-, P^{-III}, S^{2-}, Cl^-, Se^{2-} and Br^- (colored in dark blue).

As a general conclusion, weakly charged atoms form ions, but beyond +3, molecules or ions are often oxygenated or halogenated, as in the case of SO_3, SO_4^{2-}, SF_6 and ICl_4^- and the expected cation is in fact an anion.

In fact, not all possible ions exist, and it may sometimes be difficult to explain why. On the other hand, the stability of those observed is worth commenting on.

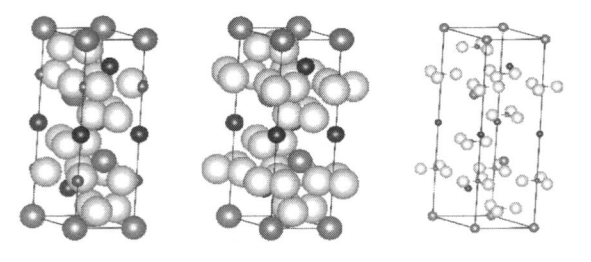

Figure 1.30. *Three-dimensional representation of dolomite $CaMg(CO_3)_2$. For a color version of this figure, see www.iste.co.uk/valls/inorganic.zip*

Certain highly charged cations do not exist in monatomic form, and we will use the term oxidation state to qualify their charge (though not a standard term, it helps to unambiguously define the described ion). For example, the sulfur ion with 6 charges does not exist (one cannot write S^{6+}) but its oxidation state is in fact +6 in the SO_4^{2-} ion or in the SO_3 molecule; therefore it will be noted $S^{+VI}O_4^{2-}$ or $S^{+VI}O_3$ or still S^{+VI}. For example, the modeling of dolomite evidences Ca^{2+} and Mg^{2+} ions and the CO_3^{2-} cation and it is difficult to separate it into four entities (see Figure 1.30).

This notation will also be used in solids for the cations involved in strongly covalent bonds (for example, the C^{+IV}-O bond).

NOTE 1.8.– The description of ions depends on the type of ions in question. There are ions in solution, in solids, in vacuum, in reactions, etc. The existence of ions is connected to their environment.

Let us consider the example of the O^{2-} anion in CaO, which in solution yields OH^- and does not remain in the form of an anion with 2 charges (the oxide is said to be basic).

The H+ ion exists in stars or in the Earth's ionosphere but when in solution it is referred to as H3O+, because it is associated with water molecules.

NO2+ can be observed in the ionosphere or in the vacuum of a spectrometer, but it is NO2– that is more stable and NO3– that is the most common.

The S+IV cation, which is virtual, for example, in the form of sulfate ions either in solution or in solids or atypical ions such as a compound of copper IV in ions such as CuF62–.

The CH3+, OH+ or CN+ ions are also formed in mass spectrometers, the NO2+, Br+, C2H7+ ions are reaction intermediates and H3+, CH2+, HCO+ ions are present in space.

From a chemist's perspective, ions are present in solution; but in this case, the size is not an essential criterion; on the contrary, the ions mentioned in Figure 1.31 relate to solids in which size plays a very important role.

1.4.2. *Radius of ions through the classification*

If ionic radii are observed by means of the classification, it can be noted that radii increase within one family, for example that of alkaline or halogen elements. Within one period, for example from Na^+ to Al^{3+}, the size of ions decreases while the charge increases, in the case of metals.

It is difficult to visualize the notion of periodicity, and the evolution of ionic radii is perceived as complex.

Several simple rules can be stated; for cations, the ionic radius is smaller than the atomic radius and it decreases when the charge increases. When the atom loses electrons, the attraction of protons in the nucleus induces compression of electron shells.

1+	2+										3+	4+	5+	2−	1−	
H																He
Li 76	Be 45										B	C 16	N 13	O 140	F 133	Ne
Na 102	Mg 72	3+	2+	2+	2+	2+	2+	2+	2+	2+	Al 54	Si 40	P 38	S 184	Cl 181	Ar
K 138	Ca 100	Sc 75	Ti 86	V 79	Cr 73	Mn 83	Fe 61	Co 65	Ni 69	Cu 77	Zn 74	Ga 62	Ge 53	As 46	Se 198	Br 196	Kr
Rb 152	Sr 118	Y 90	Zr 74	Nb 72	Mo 69	Tc	Ru 68	Rh 67	Pd 86	Ag 115	Cd 95	In 80	Sn 69	Sb 60	Te 221	I 220	Xe
Cs 167	Ba 135	Lu 86	Hf	Ta 72	W	Re	Os	Ir 68	Pt 80	Au 137	Hg 102	Tl 89	Pb 78	Bi 76	Po	At	Rn

Figure 1.31. *Size of (six-coordinate) ions in the classification. For a color version of this figure, see www.iste.co.uk/valls/inorganic.zip*

For anions, the ionic radius is larger than the atomic radius and it increases with charge, as the number of electrons exceeds that of protons in the nucleus. Moreover, repulsion between electrons increases while attraction exerted by the nucleus remains unchanged, and so electrons move away from the nucleus, leading to an expansion of the electron cloud.

If six-coordinate ions are considered the most common (see Figure 1.31), an overall observation is that negative ions (in dark blue) are larger than positive ions (with five exceptions in this figure: K^+, Rb^+, Cs^+, Au^+ and Ba^{2+} cations, indicated on medium-blue background).

A left-to-right decrease can be noted for ions, whether anions or cations (knowing that in the d-block there are only cations with 2 or 3 charges), indicated in light blue.

The smallest cation is the leftmost and uppermost one in the classification and the highest-charged are N^{3+}, C^{4+}, P^{5+}, etc. The largest anion is the rightmost and lowermost in the classification, which means Te^{2-}, then I^-, then Se^{2-}, etc.

Li$^+$, Be^{2+}, C^{4+} and N^{5+} ions have 1s^2 configuration; they are therefore considered helium isoelectronic (same number of electrons) and are indicated in descending order of size, as the nuclei have three to seven protons. The notion of isoelectronic ions will be used for ion size-comparison purposes, as, given an equivalent number of electrons, the higher the charge of nucleus, the stronger the attraction exerted by the nucleus and the smaller the ion size.

Argon isoelectronic ions are as follows: O^{2-}, F$^-$, Na$^+$, Mg^{2+} and Al^{3+}. They all have the same number of electrons for a given number of protons, which is increasing in this series, with O^{2-} being the largest cation and Al^{3+} the smallest.

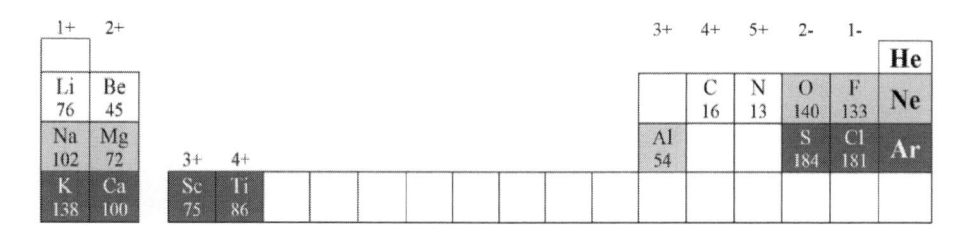

1+	2+							3+	4+	5+	2-	1-	
													He
Li 76	Be 45								C 16	N 13	O 140	F 133	Ne
Na 102	Mg 72	3+	4+					Al 54			S 184	Cl 181	Ar
K 138	Ca 100	Sc 75	Ti 86										

Figure 1.32. *Examples of isoelectronic ions. For a color version of this figure, see www.iste.co.uk/valls/inorganic.zip*

Krypton isoelectronic ions are as follows: S^{2-}, Cl$^-$, K$^+$, Ca^{2+}, Sc^{3+} and Ti^{4+}. The largest cation of this series is S^{2-} and the smallest is Ti^{4+}.

It is worth noting that cations are distributed from N^{5+} (13 pm) to Cs$^+$ (167 pm), and anions from F$^-$ (133 pm) to Te^{2-} (221 pm).

NOTE 1.9.– The fluorine ion is the smallest of anions and the cesium ion is the largest of cations. A list of cations that are larger than fluorine and of anions that are smaller than (six-coordinate) cesium can be drawn up. These are Ba^{2+}, Au$^+$, K$^+$, Tl$^+$, O^2, Ra^{2+} and Rb$^+$ ions. It can be said that there are several large-sized cations, but cations are generally smaller than anions.

Size-based ranking of ions is easy to do by drawing a classification by the number of electrons and then by the number of protons. If Ne, Ar, Li$^+$, Na$^+$, K$^+$, Mg^{2+}, Ca^{2+}, Al^{3+}, Sc^{3+}, Ti^{4+}, F$^-$, Cl$^-$, O^{2-} and S^{2-} are ranked based on the number of electrons (isoelectronic), the result is O2–, F, Ne, Na+, Mg2+ and Al3+, and then S2–, Cl–, Ar, K+, Ca2+, Sc3+ and Ti4+.

This clearly shows that while the size of the sulfur atom is 104 pm, that of the S2- anion is 184 pm (argon isoelectronic but having two protons less in the nucleus) and that of S6+ is 29 pm (neon isoelectronic but having six protons more in the nucleus).

1.4.3. *Polarizability*

Crystal polarizability influences ionic radii; therefore, it is important to memorize its evolution for all the elements.

Polarization leads to ions losing their spherical form, and the distance between them decreases. A more or less covalent bond is formed, namely the ionic covalent bond.

Polarizability can be defined by considering an atom in an electric field (generated by an ion, a molecule, etc.); its electron cloud becomes temporarily deformed.

A temporary dipole moment is induced, which is proportional to the electric field applied, with the coefficient of proportionality (α) being the polarizability.

This coefficient is all the higher as the cloud is easily deformed and it is homogeneous in a volume, with its measurement unit being generally 10^{-30} m^3 (other units are also used, such as Å^3 or 10^{-24} cm^3).

As Figure 1.33 indicates, the larger the electron cloud of the atom (or of the ion, see Figure 1.34), the higher its polarizability and the easier its deformation.

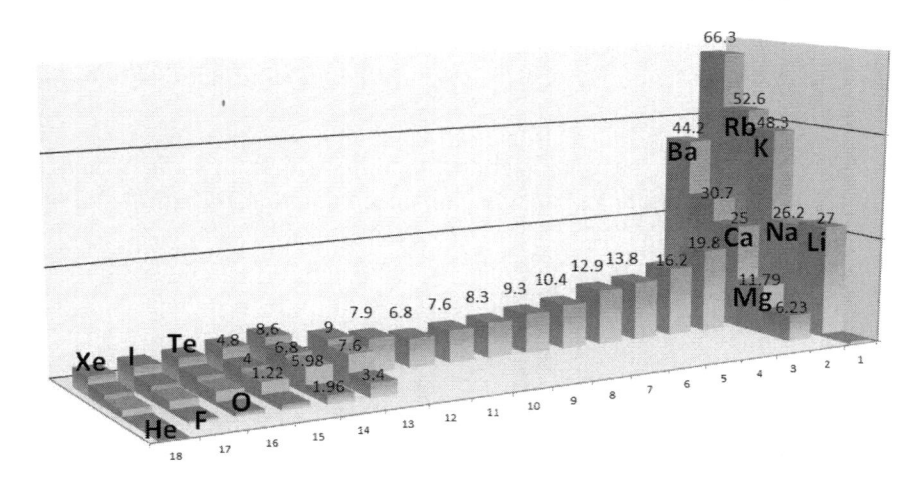

Figure 1.33. *Evolution of polarizability for the elements in the classification*

The same is valid for ions, of which the larger will be the most polarizable, though their electronegativity also plays a role, as well as the difference between the number of protons and of electrons.

The polarizability of ions increases when they are negatively charged, as their size increases and the attraction exerted by the nucleus on electrons is weaker.

The polarizability of ions is very weak when they are positively charged, as they strongly decrease in size and the attraction exerted by the nucleus on the remaining electrons is stronger (see Cs and Cs^+ in Figures 1.33 and 1.34).

Figure 1.33 shows that alkaline and alkaline-earth atoms are highly polarizable (27 for lithium to 66.3 for cesium, expressed in 10^{-30} m^3) while all the others have much lower polarizability.

Figure 1.34 shows that cations are overall much less polarizable (maximum 2.4) and the polarizability of some anions has strongly increased (for tellurium it has increased from 6 to 14 and all the anions in columns 16 and 17 have doubled their polarizability).

When halogens gain one electron and chalcogenes gain two electrons, they become larger and more polarizable; in fact, they become the most polarizable ions.

The most significant polarizabilities are exhibited by the atoms in columns 1 and 2 (12 to 66) and by the ions in columns 16 and 17 (3 to 14).

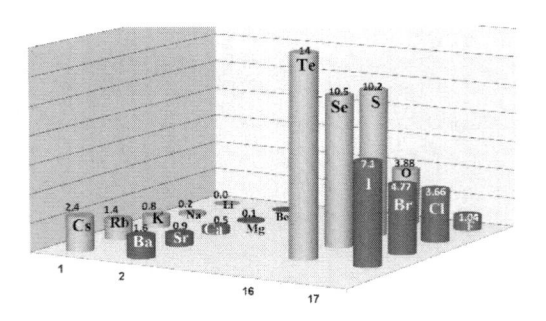

Figure 1.34. *Evolution of the polarizability of ions in the classification*

When they lose one or several electrons, cations become smaller and less polarizable, but the smaller and higher-charged ones become polarizers (for example, Si^{4+}, Ti^{4+} or Al^{3+}).

When a polarizable anion is placed next to a polarizer cation in a crystal, the two effects add up and the bond is strongly covalent.

1.4.4. *The radii of ions in solids*

The evolution of the size of six-coordinate ions through the classification has been described without mentioning how the sizes, which vary strongly depending on various parameters, have been obtained.

The size of ions in solids is generally deduced from the experimental distance between neighboring cations and anions, as determined by X-ray analysis. It relies on geometrical considerations, and therefore depends on the size of the ion taken as reference, which is generally O^{2-} or F^-.

As readily noted, valence, coordination number, and the nature of the bond that structures the ions, which is mainly given by their electronegativity and polarizability, influence their size (spin state, temperature and pressure also have an impact).

Figures 1.35 and 1.36 propose several examples that illustrate the effect of these factors on the size of ions. The hard spheres model, which is well-adapted to metals, could not be generalized as such to all crystalline compounds.

Other methods have therefore been used in order to obtain absolute ionic radii from electron density maps [MEI 67] or physical constants such as molecular refraction [WAS 23], or from the thermal expansion coefficient [FUM 64], unfortunately with diverging results.

In agreement with P. Besançon [BES 82], we propose that the physical and geometric scales reflect different realities, and therefore it is useless to try to merge them.

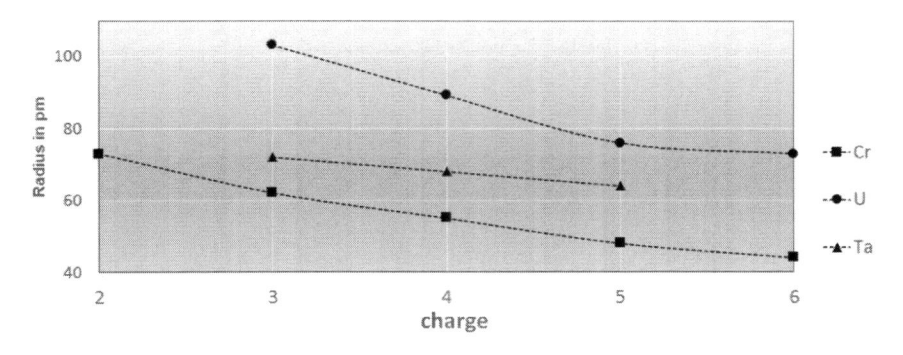

Figure 1.35. *Ionic radius of several ions depending on their valence*

To summarize the set of observations, it can be proposed [BES 82] that for cations there is a negative correlation between the ionic radius and valence, but a positive correlation between ionicity and the coordination number. For anions, there is a positive correlation between the ionic radius and valence, as well as for the coordination number, but with a more moderate influence than for cations.

Finally, for ions there is a positive correlation between the coordination number and ionicity.

No simple law has been proposed that would predict the ionic radius; therefore the simplest models already described, such as the perfect crystal model and the hard spheres model, will be used.

Nevertheless, it will be considered that ions are composed of two parts: one hard central nucleus that concentrates most of the electron density, and an outer soft sphere, more or less polarizable, whose electron density is lower.

When conditions are set for an ion, and an ionic radius is assigned to it, it is easy to imagine a hard sphere and we get back to the previously described diagrams.

It can be proposed that anions and cations are in contact in the crystalline solid and any gap, instead of invalidating the models, will rather serve to detect the contribution of covalence to the bond.

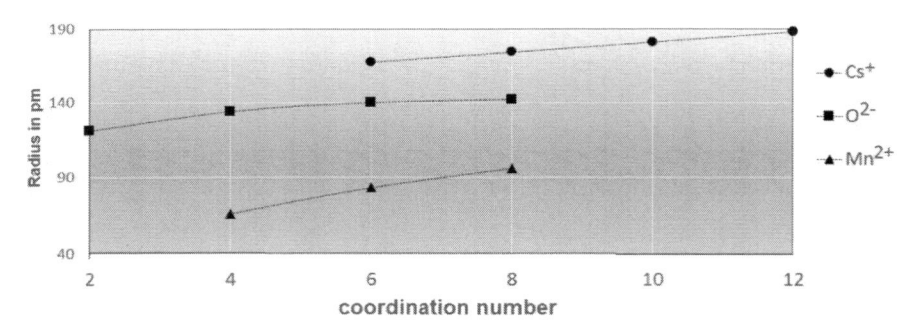

Figure 1.36. *Ionic radius of several ions depending on their coordination number*

The limits of these two models will be exceeded even further as the type of bond evolves from ionic to covalent (with strong electron cloud overlap and the polarization of ions that deforms the electron clouds) or as it evolves to metallic character.

When there is a significant gap, other criteria, such as crystal color or lack of solubility, confirm this evolution. For example, NiAs is bright metallic black and insoluble in water and certainly is not an ionic crystal.

Based on these propositions, ions can be assimilated to hard spheres that are in contact with one another, as long as their valence and coordination number (and their spin state, if needed) are fixed.

Ag^+	Ag^{2+}	Al^{3+}	As^{3+}	As^{5+}	Au^+	Au^{3+}	Ba^{2+}	Be^{2+}	Bi^{3+}	Bi^{5+}
115	94	54	58	46	137	85	135	45	103	76
Br^-	Br^{7+}	Ca^{2+}	Cd^{2+}	Ce^{3+}	Ce^{4+}	Cl^-	Co^{2+}	Co^{3+}	Cr^{2+}	Cr^{3+}
196	39	100	95	101	87	181	65	55	73	62
Cr^{4+}	Cr^{6+}	Cs^+	Cu^+	F^-	F^{7+}	Fe^{2+}	Fe^{3+}	Ga^{3+}	Gd^{3+}	Ge^{2+}
55	44	167	77	133	8	61	55	62	94	73
Ge^{4+}	Hf^{4+}	Hg^+	Hg^{2+}	I^{5+}	I^{7+}	In^{3+}	In^{5+}	Ir^{3+}	Ir^{4+}	K^+
53	71	119	102	95	53	80	57	68	63	138
La^{3+}	Li^+	Lu^{3+}	Mg^{2+}	Mn^{2+}	Mn^{3+}	Mn^{4+}	Mo^{3+}	Mo^{4+}	Mo^{5+}	N^{3+}
103	76	86	72	83	58	53	69	65	61	16
N^{5+}	Na^+	Nb^{3+}	Nb^{4+}	Nb^{5+}	Ni^{2+}	Ni^{3+}	O^{2-}	OH^-	Os^{4+}	Os^{5+}
13	102	72	68	64	69	56	140	137	63	58
P^{5+}	Pb^{2+}	Pb^{4+}	Pd^{2+}	Pd^{3+}	Pd^{4+}	Po^{+4}	Pt^{2+}	Pt^{4+}	Pu^{+3}	Pu^{+4}
38	119	78	86	76	62	97	80	63	100	86
Pu^{+6}	Rb^+	Re^{+4}	Re^{+5}	Re^{+6}	Re^{+7}	S^{2-}	S^{4+}	S^{6+}	Sb^{3+}	Sb^{5+}
71	152	63	58	55	53	184	37	29	76	60
Sc^{3+}	Se^{2-}	Se^{4+}	Se^{6+}	Si^{4+}	Sn^{4+}	Sr^{2+}	Ta^{+3}	Ta^{+4}	Ta^{+5}	Tc^{+4}
75	198	50	42	40	69	118	72	68	64	65
Te^{2-}	Te^{4+}	Te^{6+}	Ti^{2+}	Ti^{3+}	Ti^{4+}	Tl^+	Tl^{3+}	U^{3+}	U^{4+}	U^{5+}
221	97	56	86	67	61	150	89	103	89	76
U^{6+}	V^{2+}	V^{3+}	V^{4+}	V^{5+}	W^{4+}	W^{6+}	Y^{3+}	Yb^{2+}	Zn^{2+}	Zr^{4+}
73	79	64	58	54	66	60	90	102	74	72

Figure 1.37. *Ionic radius of the 6-coordinate ions mentioned [HAY 15]*

NOTE 1.10.– In order to avoid confusion, non-standardized nomenclature will be used, so that the ion coordination number can be undoubtedly qualified, by indicating between parentheses, and in roman numerals, the number of neighbors of opposite charge. For example, the calcium ion whose coordination number can be 6, 8, 10 and 12 will be written as $Ca^{2+(VI)}$, $Ca^{2+(VIII)}$, $Ca^{2+(X)}$ and $Ca^{2+(XII)}$.

Based on these propositions, a good approximation is to consider two categories of ionic radii:

– Anions with 1 charge and whose size varies little, such as oxygen $O^{2-(II)}$ –121 pm to $O^{2-(VIII)}$ –142 pm (15% variation), or which can be generally considered independent of the coordination number, and which is given in six-coordinate: F^- (133 pm), Cl^- (181 pm), Br^- (196 pm), I^- (220 pm), S^{2-} (184 pm), Se^{2-} (198 pm) and Te^{2-} (221 pm);

– Cations whose radius depends mainly on charge, coordination number and, for some, on spin state (chromium, cobalt, manganese and iron ions with 2 and 3 charges, as well as nickel ion with 3 charges).

For example, the size of lead cations can double if their charge and coordination number evolve: $Pb^{2+(XII)}$ has an ionic radius of 149 pm, $Pb^{2+(VI)}$ of 119 pm, $Pb^{4+(VIII)}$ of 94 pm and $Pb^{4+(VI)}$ of 78 pm.

The values used in this book are given in Figure 1.37, and they are proposed in the Handbook [HAY 15].

Before using these data, it is worth noting that chemistry is preeminently an experimental science and that ion radii are set by their coordination number and based on various measurements conducted in particular by Shannon et al. [SHA 69] and on calculations made by Pauling [PAU 60]. There is no available theory that allows the *a priori* determination of ionic radii, and the use of the proposed values should take into consideration that these are average values.

Likewise, electronegativities are provided for each element irrespective of its valence or environment; therefore informed use is recommended.

What then is the interest of the proposal, it may be asked. Perfectly correlated data are impossible to obtain other than on a case-by-case basis and it only leads to complexity, which cannot contribute to a simple overall view on the solid.

The criteria employed are calculated on the basis of electronegativity and ionic radii, and careful use is therefore recommended. Tendencies, correlations or simple relations will be evidenced, but no rigorous laws will be highlighted. The objective is to offer several keys that help in qualifying a crystal and in understanding its overall structuring, as well as its evolution in series of compounds based on experimental data. The approach is willingly empirical, giving the reader the freedom to go beyond what is proposed and apply it in his specific field.

2

Knowledge of Metallic Crystals

2.1. Properties of metals

Solid matter is composed of atoms, ions or molecules that ideally get assembled in an ordered manner and form a crystal, which is often visualized as gemstones, although metals, ice or iodine can also take crystalline form.

Since crystals are effectively part of the mineral kingdom and cover only a limited number of structures commonly shown in the mineralogy literature, the crystalline form of gases or liquids that are cooled until solidification can also be studied.

Any substance under a certain temperature becomes solid (or liquid, as intended); thus, an adjusted value of this temperature separates the solid and liquid states. Moreover, determination of this temperature may prove difficult even for metals, since tungsten is liquid at 3422°C, whereas mercury at −38.8°C, for example.

Current technologies require ever more materials with specific characteristics, and a good understanding of solid states and their properties led to the development of radar-absorbent materials (stealth aircraft), piezoelectric materials (loudspeakers) and superconductors (medical imaging).

A large part of the objects in our environment are in crystalline form (metals, oxides, ceramics, minerals, salts, polymers, etc.). The first to be studied here are pure metals, their packing being among the simplest, due to the identical nature and size of their atoms, which are electrically neutral.

2.1.1. *Characteristics of the metallic bond*

A preliminary step in the description of how metals are packed is to understand what keeps metal atoms together, as the nature of the metallic bond will prove useful further on, when studying metal packing.

Metal sublimation energy has the same order of magnitude as covalent bond dissociation energy (several hundred kilojoules per mole), which is characteristic of a strong bond. Instead of being shared between two atoms, electrons are shared among all atoms. The metallic bond is therefore strong and non-directional or isotropic, which explains the properties of metals and, in particular, their preferential packing of atoms, which have as many neighbors as possible, or their ability to become deformed without breaking.

Unlike the covalent bond, the metallic bond is delocalized over a significant number of atoms. Indeed, the number of valence electrons is not sufficient for the formation of a bond between pairs of atoms (each atom is generally surrounded by 8 or 12 neighbors), each of which can be considered as an ion immersed in a gas of electrons belonging to a very significant number of atoms, the whole being overall neutral.

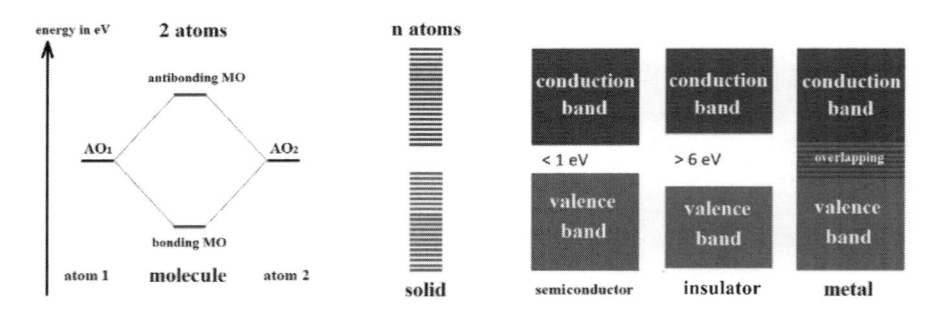

Figure 2.1. *Simplified diagram of the energy-band model. For a color version of this figure, see www.iste.co.uk/valls/inorganic.zip*

In an isolated atom, electrons occupy specific and distinct levels of energy in their atomic orbitals. According to the Pauli exclusion principle, the maximum number of electrons that may have the same energy is two. Moreover, when two atoms form a covalent bond, the Pauli exclusion principle would be violated if the energy levels of the two electrons forming the bond were identical. Their energy levels split into two different energy levels, which are qualified as bonding and anti-bonding.

The same behavior can be observed when two atoms come together to form a solid; however, when the number of atoms increases, these two levels also split into other neighboring energy levels. When the number of atoms becomes very significant, a series of molecular orbitals are formed, whose energy varies in a practically continuous manner, and they are said to constitute an energy band. This phenomenon is described by the energy-band model.

The last filled band is called the valence band and the conduction band is immediately above it (see Figure 2.1). These two energy bands can overlap (metals) or be separated by a forbidden band also called gap, whose energy value dictates the electrical properties of the material:

– In the case of conductors, and therefore metals, the valence band and the conduction band overlap and hence the electrons can migrate to the conduction band and move throughout the solid, thereby rendering the metal a good conductor.

– For semiconductors, the valence band and the conduction band are close and the gap is close to 1 eV. Conduction is weak at ambient temperature, but rapidly increases with temperature, and when electrons receive energy (by heating, application of an electromagnetic field or by exposure to light), some of them are able to move into the conduction band and in the material.

– As for the insulators, the valence band and the conduction band are separated by a so-called forbidden band of approximately 6 eV. This is too high a value for the electrons to be able to move into the conduction band without gaining a significant amount of energy.

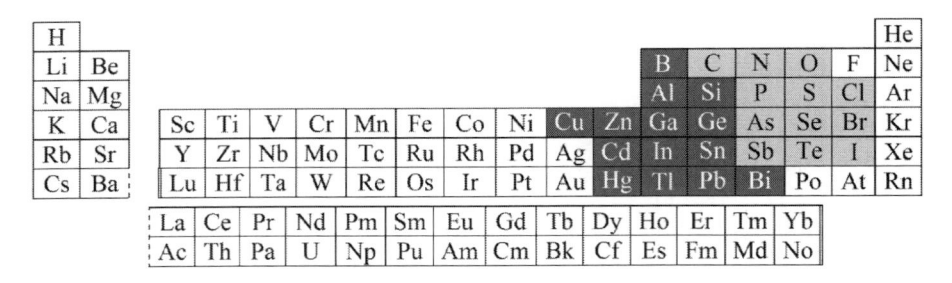

H																	He
Li	Be											B	C	N	O	F	Ne
Na	Mg											Al	Si	P	S	Cl	Ar
K	Ca	Sc	Ti	V	Cr	Mn	Fe	Co	Ni	Cu	Zn	Ga	Ge	As	Se	Br	Kr
Rb	Sr	Y	Zr	Nb	Mo	Tc	Ru	Rh	Pd	Ag	Cd	In	Sn	Sb	Te	I	Xe
Cs	Ba	Lu	Hf	Ta	W	Re	Os	Ir	Pt	Au	Hg	Tl	Pb	Bi	Po	At	Rn

La	Ce	Pr	Nd	Pm	Sm	Eu	Gd	Tb	Dy	Ho	Er	Tm	Yb
Ac	Th	Pa	U	Np	Pu	Am	Cm	Bk	Cf	Es	Fm	Md	No

Figure 2.2. *Distribution of common semiconductors in the classification. For a color version of this figure, see www.iste.co.uk/valls/inorganic.zip*

Semiconductors [HAY 15] occupy columns 11–15, which are marked by dark blue, and several metals of d- and f-blocks (titanium, molybdenum, platinum and uranium) as well as carbides, nitrides, phosphides, oxides, chlorides, selenides, etc., resulting from the combination with elements highlighted in light blue (see Figure 2.2).

2.1.2. *Conductivity and the melting temperature of elements*

2.1.2.1. *Conductivity*

The electrical conductivity of elements can be visualized through the classification but it only presents an interest in the case of metals.

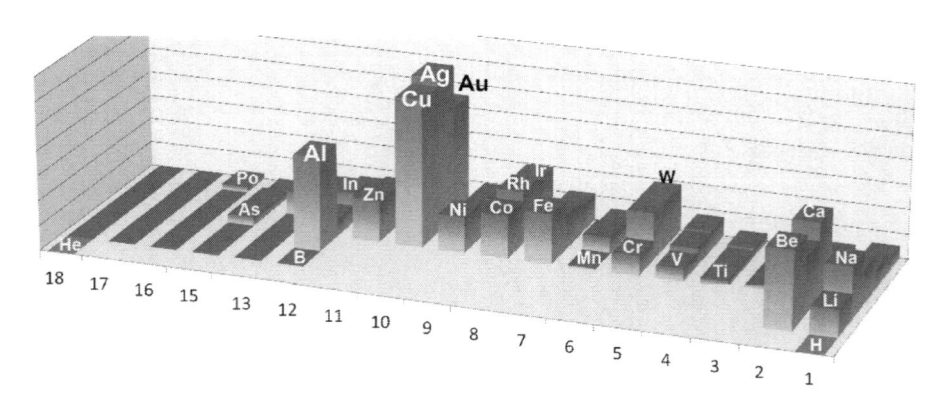

Figure 2.3. *Conductivity of elements*

Metals have generally significant electrical and thermal conductivities. These are due to the electron cloud in which atoms are immersed, as the agitation induced by an electric field or by heat transfer is easily propagated by these electrons, and contributes to electrical or thermal conduction.

For better visibility, the classification shown in Figure 2.3 is reversed, the first period being in the front, while the sixth period is in the back.

For simplification purposes, metals can be generally considered as good electric and thermal conductors; taking copper conductivity as a reference, the electrical conductivity of aluminum is 1.6 times smaller, that of iron is 3 times smaller, that of titanium is 25 times smaller and that of manganese is 86 times smaller. Moreover, all metals have a shiny silver brightness (except copper and gold); they are solid at ambient temperature (except mercury) and have good malleability and ductility.

It is useful to note that while all metals are conductive, some of them are very weak conductors; for example, manganese and plutonium are approximately 100 times less conductive than copper, whereas more than 3,000 times more conductive than tellurium, which is the most conductive non-metal. To give an idea of the order of magnitude, electrical conductivities of copper and manganese are $6.3.10^7$ and $6.9.10^5$ S/m, respectively.

Through all the periods of the periodic table, except for the first one, the elements situated on the far left side are good conductors, while those on the far right side are very weak conductors. The d-block with noble metals contains elements whose conductivities are the highest.

> NOTE 2.1.– Let us note that it is difficult to replace copper as electrical conductor with another metal of comparable conductivity, knowing that silver and gold are rare (800 and 15,000 times rarer, respectively).

> Alkaline and alkaline-earth elements are too reactive, so aluminum is the only possible replacement; however, it should also be taken into account that it is 1.6 times less conductive. Moreover, it is the third most abundant element of the Earth's crust (1,400 times more available in the Earth's crust than copper). Unfortunately, its production is highly energy consuming as its oxide is very stable and it is difficult to obtain the metal by classical methods, electrolysis of molten salts being

used instead. Antimony and arsenic are metalloids but unlike their homologues, they are very average conductors (about 20 times less than copper), to the same extent as barium or titanium.

2.1.2.2. Melting temperature

Melting temperatures are higher for metallic elements than for non-metallic elements in the periodic table.

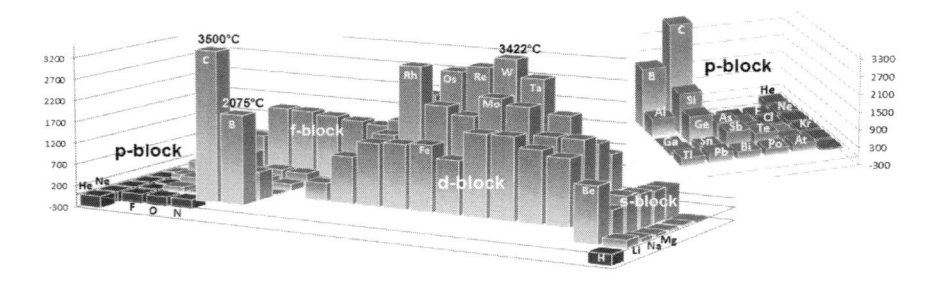

Figure 2.4. *Melting temperatures of elements*

The elements with the highest melting points are boron in metalloids, carbon in non-metals and tungsten in metals. Elements with melting points above 2,000°C are qualified as refractory.

Similar to conductivity, melting temperatures vary strongly for metals (see Figure 2.4); however, this is valid only for a part of non-metals. The most refractory is the carbon used in Thomas Edison's incandescent light bulb, followed by its logical replacement, tungsten.

> NOTE 2.2.– Non-metals exhibit the lowest temperatures, a very particular example being helium, which exists in solid state only below 0.95 K and becomes a gas at 4.2 K.

> Metals also include some extreme elements: mercury exists in liquid state from −38.8°C and is therefore liquid at ambient temperature. This is not surprising and, due to its properties, mercury has many applications in everyday life, but it is not the only metal that can be found in liquid state at ambient temperature. Francium (27°C), gallium (29°C) and rubidium (39°C) have melting points significantly above the melting point of mercury!

2.2. Study of packing in metals

Metallic crystals or more simply called metals are formed by a juxtaposition of identical atoms united by a metallic bond. This is an isotropic (non-directional) bond and each atom has generally 8 or 12 neighbors.

In order to describe metals, the atoms are assimilated to hard, non-deformable spheres of given diameter, which are in contact with one another along certain directions. It is the hard sphere model that is used to facilitate the description of metal properties.

Metals have a crystalline structure due to which they can be considered an assembly of identical spheres. A description of how the atoms are packed in a metal, as well as of the parameters that characterize these packings, is the focus of this section.

In order to draw a simple representation that will be designated as a perfect crystal model, the crystal is considered infinitely identical and therefore devoid of any defects. This is an ideal model, since real crystals always have defects, which confer them often useful properties.

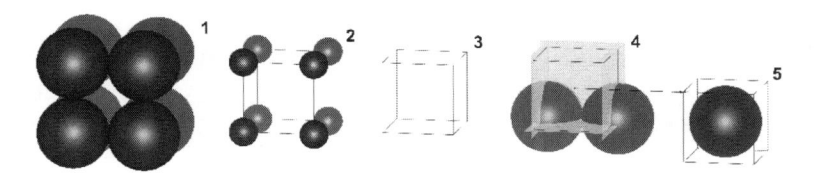

Figure 2.5. *Representation of primitive cubic*

Once these two models have been defined, it is possible to represent metals as a perfect packing of hard spheres (see (1) in Figure 2.5).

The centers of the eight atoms define a cube that is not clearly visible; therefore, packings are generally represented by small balls united by lines (2), which give the possibility of viewing the structure formed by the centers of the eight atoms, named the unit cell (3), which in this case is a cube. Infinite repetition of this type of cube or unit cell fills the whole space and yields the crystal [ANG 01].

Inside the unit cell, there is only a fraction of the atoms represented, and visualization (4) in the figure only includes one-eighth of each atom inside it.

Considering that eight times one-eighth of an atom means that one atom is present in the unit cell, a better visualization can be obtained by moving the top of the unit cell (3) to the center of the previous one (5). In this case, the central atom is fully contained in this new unit cell (one atom can be effectively counted in the unit cell) and it is tangent to the six faces of the cube representing it (5).

However, this representation does not allow the visualization of the unit cell and thus the previous representation with eight atoms is always preferred.

A crystalline structure can be described as periodic and infinite repetition throughout the space of identical volumes, that is, of unit cells (3). It is represented in various formats depending on whether what should be observed are the atoms in contact (1), the geometry of the unit cell (2) or the number of atoms contained (5).

2.2.1. *Formation of planar packing*

The hard sphere model allows us to propose the format of ping-pong balls (perfectly uniform in size and non-deformable), for example, in order to provide a description of the packing of atoms in a metal.

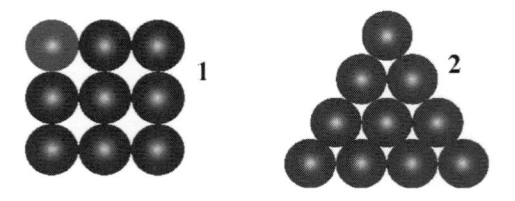

Figure 2.6. *Planar non-close-packed (N) and close-packed (A) structures*

Figure 2.6 presents two types of planar packing: non-close-packed (denoted by N) and close-packed (denoted by A). The packing that features more empty space is designated as non-close-packed (1), while the other one (2) is called close-packed. These two types of packing correspond to two ways of regular planar packing of spheres of the same radius so that the least possible space is left between them.

If the reader does not find the above obvious, he/she is invited to conduct the following experiment:

– Randomly place one layer of ping-pong balls in a shallow box; the result will be a random packing (1) (see Figure 2.7);

Figure 2.7. *Random planar non-close-packed (N) and close-packed (A) structures. For a color version of this figure, see www.iste.co.uk/valls/inorganic.zip*

– Gently shake the box and slightly tilt it to one corner so that the balls form a packing N (2), being guided by the walls of the box;

– Shake the box once again but tilt it toward you so that only one ball is in contact with the left wall (for example) of the box; you will note a quite high packing A (3).

The arbitrary packing (1) of the balls has uneven (enclosed) spaces that are larger than the ones in the other packings (2 and 3) in which these spaces are perfectly even.

The non-close-packed structure (2) reduces the size of these spaces, and the close-packed packing (3) reduces it even more. The closest-packed structure is possible in a plane. It should be noted in this experiment that the energy is minimized by gravity, which brings the balls as close to one another as possible due to their mass.

2.2.2. *Crystal formation*

The formation of a three-dimensional crystal requires the packing of planes, which can be achieved in different ways depending on the initial planar type (N or A).

2.2.2.1. *Packing of N planes*

The packing of N planes represented in Figure 2.8 (1 and 2) yields only one model (3) called cubic P (primitive cubic or simple cubic).

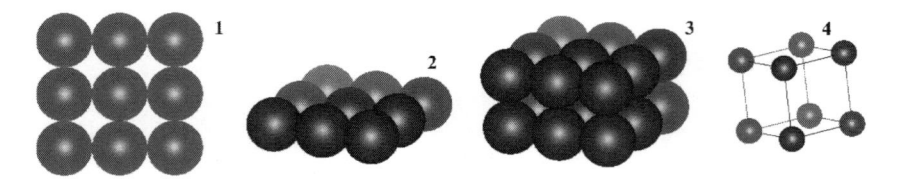

Figure 2.8. *Packing of N-type planes*

The centers of eight arbitrary neighboring atoms define a cube, which can be visualized by reducing the size of the atoms represented and connecting them through lines to highlight the unit cell (4).

NOTE 2.3.– NN' packing means packing of two N planes so that the second one (N') fits in the hollows of the first one (see figure below); this type of packing corresponds to a centered tile of square basis (1 and 2).

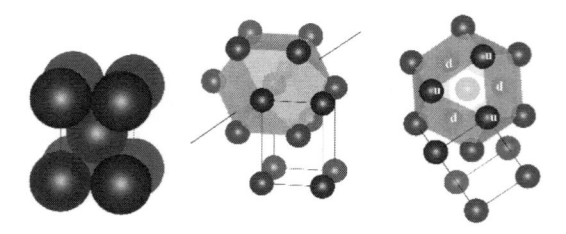

Considering several atoms of neighboring unit cells, a cuboctahedron (3) appears, which has been repositioned (4) according to the highlighted axis in order to visualize the triangles shifted upward (u) and downward (d) of the cuboctahedron.

It can therefore be concluded that NN' packing is in fact an ABC packing (see Figure 2.11).

2.2.2.2. *Packing of A planes*

In order to define the types of packing in metals, the first plane (and any identical plane) is denoted by A, the plane above it, if shifted, is denoted by B, and the third one, which is shifted compared to the first two, is denoted by C (only three types of planes are possible).

The packing of A planes leads to two models, as the packing of AA planes, similarly to that of NN planes, is not observed in metals (it would be a hexagonal packing, see Figure 2.10).

The first packing of A planes is necessarily of AB type (A and B are similar, but shifted), and this notion of shift will prove useful later on.

Figure 2.9 presents sites 1 and 2 at which an atom can be placed. It should be noted that one type cannot be occupied unless the other one is free, which is due to the size of atoms. Moreover, as soon as one atom is placed, the set of positions of the remaining atoms on that level is fixed.

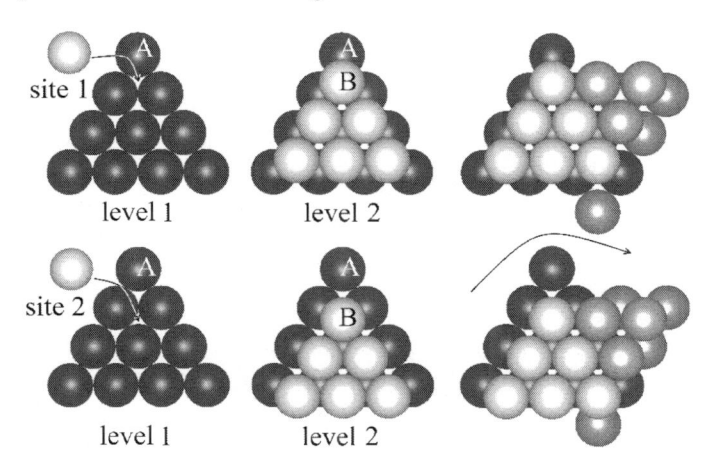

Figure 2.9. *Packing of two A-planes. For a color version of this figure, see www.iste.co.uk/valls/inorganic.zip*

Regardless of the choice for the first site, the same AB packing is obtained, and this is highlighted by the addition of several atoms (in gray) followed by a 180° rotation (marked by an arrow in Figure 2.9).

It is the third level that makes the difference between two packings, since while the choice of either site 1 or 2 for the first level has no influence, the choice of site 1 or 2 for the second level leads to two different types.

If sites 2 of the second level are chosen, a further A plane is created, and the result will be an ABA type of packing or more simply an AB type, due to the alternate succession of A and B planes (see Figure 2.10).

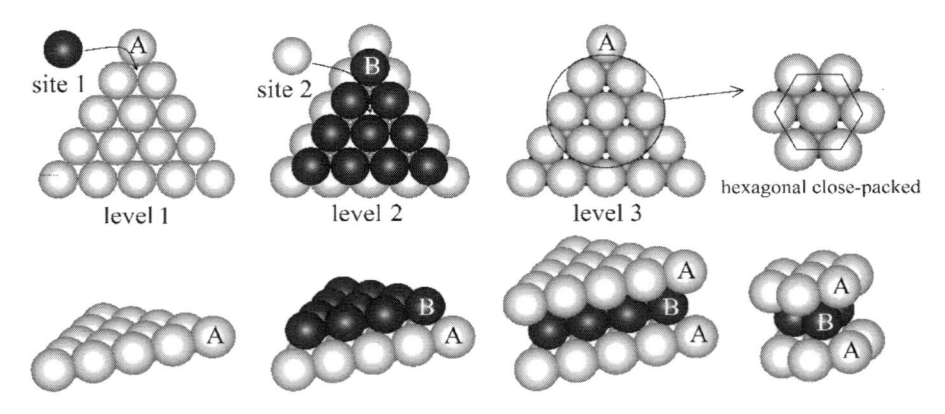

Figure 2.10. *AB packing of three A-type planes. For a color version of this figure, see www.iste.co.uk/valls/inorganic.zip*

This packing can be associated with a hexagon, the hexagonal close-packed (hcp) or AB packing.

The definition of the unit cell in this type of packing will be detailed further on, as it does not contribute to a better understanding of the descriptions in this section. It should however be noted that the volume associated with AB packing is the hexagon.

The second packing of A planes is of ABC type, where A, B and C planes are similar but shifted (see Figure 2.11).

A convenient choice of atoms highlights the cube whose faces are all centered. Conversely, in a face-centered cube, an ABC packing can be highlighted (see Figure 2.12). This packing is of ABC type, all face-centered cubic (fcc), cubic F or cubic close-packed.

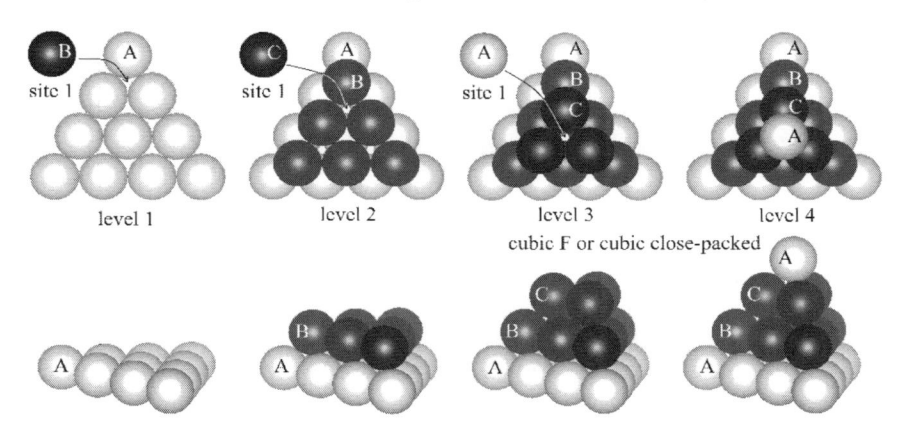

Figure 2.11. *ABC packing of three A-type planes. For a color version of this figure, see www.iste.co.uk/valls/inorganic.zip*

If sites 1 of the second level are chosen, a plane C is created and the result is an ABC-type of packing, which is formed of an infinite alternate series of A, B and C planes.

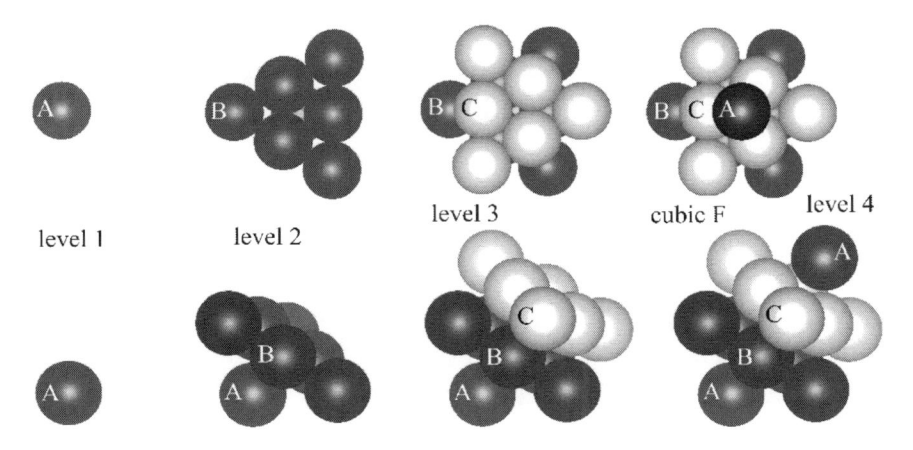

Figure 2.12. *ABC packing of three A-type planes. For a color version of this figure, see www.iste.co.uk/valls/inorganic.zip*

These two types of packing differ by the succession of planes, and the resulting volume is a cube (for ABC packing) or a hexagon (for AB packing).

A third type of packing present in metals is the body-centered cubic structure (1) or cubic I (see Figure 2.13). In this type of packing, the atoms are not that close (2), as in the previous two but it is a common occurrence.

NOTE 2.4.– The description of various types of packing offers simple keys for identifying them, regardless of the axis of representation.

An ABC or cubic F is visualized as either a regular succession of three close-packed layers or the presence of a cuboctahedron (1) represented in the figure below. An AB or hexagonal close-packed structure is visualized as either a regular alternation of two close-packed layers or the presence of an anti-cuboctahedron (2).

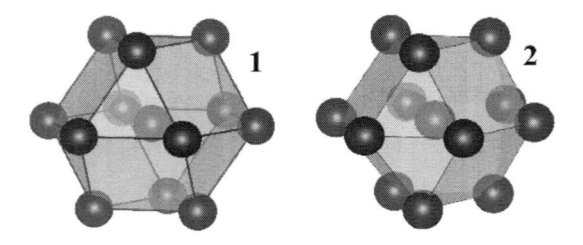

It should be noted that the reverse is verified in both cases, as cubic F is also an ABC packing and it forms cuboctahedrons, and the hexagonal close-packed structure is also an AB packing and it forms anti-cuboctahedrons.

It is a matter of alternation of non-close-packed planes SS', the atoms of each layer being separated by a distance of 0.309 times the radius of atoms, while inter-layer atoms are in contact (3).

The structure is overall less close-packed than the previous one, as its packing density is 0.68 instead of 0.74 (see section 2.2.4).

Figure 2.13. *Packing of non-close-packed planes into cubic I. For a color version of this figure, see www.iste.co.uk/valls/inorganic.zip*

If the three types of packing present in the metals are represented in a classification, there are 33 elements of hexagonal close-packed type (medium blue), 40 of cubic I (dark blue) and 28 of cubic F (light blue) that refer to metals and rare gases (see Figure 2.14). Non-metals (except for rare gases) yield diversified types that have not been mentioned here [HAY 15]. Only the diamond packing is evidenced for column 14 (except for lead).

H																	He
Li	Be											B	C^d	N	O	F	Ne
Na	Mg											Al	Si^d	P	S	Cl	Ar
K	Ca	Sc	Ti	V	Cr	Mn	Fe	Co	Ni	Cu	Zn	Ga	Ge^d	As	Se	Br	Kr
Rb	Sr	Y	Zr	Nb	Mo	Tc	Ru	Rh	Pd	Ag	Cd	In	Sn^d	Sb	Te	I	Xe
Cs	Ba	Lu	Hf	Ta	W	Re	Os	Ir	Pt	Au	Hg	Tl	Pb	Bi	Po	At	Rn
Fr	Ra																
		La	Ce	Pr	Nd	Pm	Sm	Eu	Gd	Tb	Dy	Ho	Er	Tm	Yb		
		Ac	Th	Pa	U	Np	Pu										

Figure 2.14. *Main types of packing in metals. For a color version of this figure, see www.iste.co.uk/valls/inorganic.zip*

Some metals may have various types of packing depending on temperature; for example, lithium, iron and thallium present all three types.

It can therefore be considered that metals form mainly cubic I or cubic F or hexagonal close-packed structures, as shown in Figure 2.15.

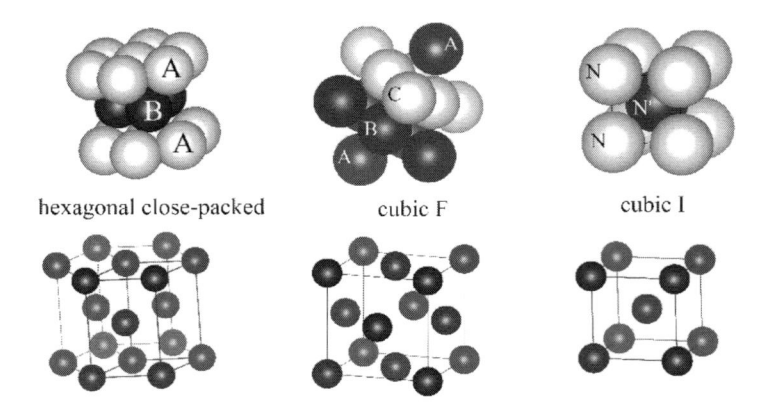

Figure 2.15. *Three main types of packing in metals. For a color version of this figure, see www.iste.co.uk/valls/inorganic.zip*

2.2.2.3. *High period packings*

Some structures are formed of dense and periodical planes of type AB/AB/... or ABC/ABC/..., but more complex structures of type ABAC/... for La, Pr, Nd and Am or of type ABACACBCB/... for Sm can sometimes be observed, and they all have the same packing density as they are close packings of planes that are in their turn close-packed. This is referred to as polytypism characterized by a sequence of high period packings. Random packings of close-packed planes are also observed but they correspond to amorphous materials and not to crystalline materials.

2.2.3. *Counting atoms in a unit cell*

It is very useful to count the atoms in a unit cell and it can be readily memorized. For a cubic unit cell, there are five most common positions (see Figure 2.16).

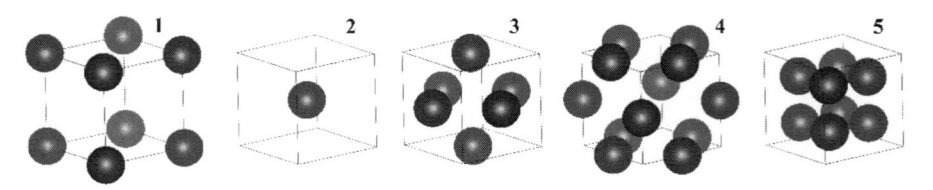

Figure 2.16. *Counting atoms in a cubic unit cell*

The eight atoms in (1) each have one-eighth in the unit cell, which amounts to one atom per unit cell. The atom in (2) and the eight atoms in (5) are entirely in the unit cell, whereas the six atoms in (3) are half in the unit cell, which amounts to three atoms in the unit cell. Finally, the 12 atoms in (4) each have one-fourth in the unit cell; hence, there are three atoms in the unit cell.

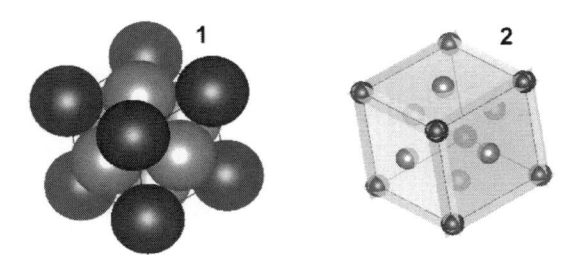

Figure 2.17. *Fractions of atoms present in the cubic F unit cell. For a color version of this figure, see www.iste.co.uk/valls/inorganic.zip*

For example, for the all-face-centered cubic structure represented by means of 14 atoms (see Figure 2.17), the atoms can be divided into two groups, eight located at the cube vertices (dark blue) and six at the centers of the six faces (in light blue):

– At each of the eight vertices, only one-eighth of an atom is in the unit cell, so there is one atom in the unit cell;

– On each of the six faces, only half of an atom is present in the unit cell, so there are three atoms in the unit cell;

Therefore, the unit cell counts one atom (corresponding to the eight vertices) plus three atoms (corresponding to the six faces), which yields four atoms.

In order to confirm the calculation (see Figure 2.18), a representation can be drawn on the basis of the cubic F structure (1), a unit cell that is shifted by 1/4 of the diagonal forward and to the left (2).

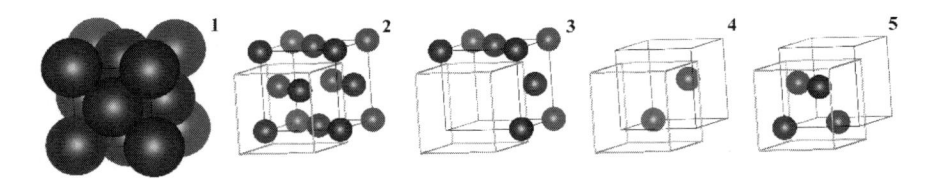

Figure 2.18. *Atoms present in a cubic F unit cell*

This representation highlights the eight atoms (3) above the new unit cell and to its right and the two atoms behind it (4). This new cube contains only four atoms (5) of the initial unit cell and all the other atoms are outside of it.

This approach can be generalized to other types of unit cell, for example, to the hexagonal close-packed structure (1 and 2). The top view (3) highlights three types of atoms (X, Y and Z) whose fraction in the unit cell is visualized (see Figure 2.19).

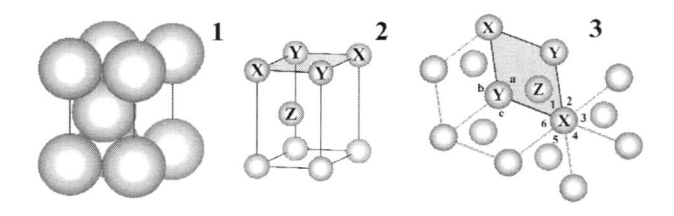

Figure 2.19. *Atoms present in the hexagonal close-packed unit cell*

The X-type is entirely in the unit cell, the Y-type belongs to three unit cells (a, b and c) and to the three unit cells above, which means that 1/6 of the atom is in the unit cell, and finally the Z-type belongs to six unit cells (1 to 6) and to six unit cells that are above, which means that 1/12 of the atom is in the unit cell. There is one atom of X-type, four atoms of Y-type and four atoms of Z-type per unit cell, amounting to $1 + 4 \times 1/6 + 4 \times 1/12$, that is, two atoms per unit cell.

> NOTE 2.5.– If the focus is exclusively on the atoms located at the eight vertices of an arbitrary unit cell, then generalization leads to the conclusion that regardless of the volume considered, if the opposite faces are parallel, the unit cell contains one atom, knowing that the fractions present in the unit cell vary, but their sum is always one atom.

2.2.4. Packing density

In the hard sphere model, the atomic packing factor or the packing density is the fraction of volume of the unit cell that is occupied by the volume of atoms (assimilated to spheres) contained in that unit cell.

It has been observed that the stackings in the plane were more or less close-packed; therefore, the notion of packing density has been extended to planes (surface packing density) and even to lines (linear packing density).

2.2.4.1. Packing density of primitive cubic structure

The lattice parameter a is defined as the length of the edge of the cube formed by the centers of eight atoms and the radius r of an atom.

For the primitive cubic structure, it can be seen in Figure 2.20 that the unit cell volume is equal to that of a cube of edge a, that is, a^3.

The volume of n atoms is calculated by the formula

$$V_{atoms} = n \frac{4}{3} \pi r^3 \qquad [2.1]$$

The relationship between a and r is determined, knowing that the atoms are in contact along one side of the unit cell and therefore the lattice parameter a is equal to 2r (1) for a number of atoms n in the unit cell that is equal to 1 (see section 2.2.3).

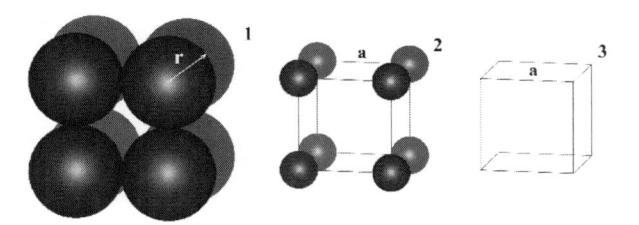

Figure 2.20. *Representations of a primitive cubic system. For a color version of this figure, see www.iste.co.uk/valls/inorganic.zip*

Packing density or fraction (f) is given by the expression

$$f = n \frac{4}{3} \pi \frac{r^3}{a^3} \qquad [2.2]$$

If a is replaced by 2r and expression [2.2] is simplified, then the result is $\frac{\pi}{6}$ or 0.52.

In the cubic P packing, atoms occupy 52% of the space and the remaining 48% is empty space between atoms.

2.2.4.2. Packing density for cubic F system

For the cubic F packing (see Figure 2.21), the number of atoms n in the unit cell is 4 (see section 2.2.3).

The relationship between a and r is obvious, since atoms are in contact along the diagonal of the face whose length is $a\sqrt{2}$ or 4r, which yields

$$a = 2r\sqrt{2} \qquad [2.3]$$

The packing density is calculated from [2.3] as $4\frac{4}{3}\pi\frac{r^3}{a^3}$ or $4\frac{4}{3}\pi\frac{r^3}{(2r\sqrt{2})^3}$

and after simplification, the result is $\pi\frac{\sqrt{2}}{6}$ or $\frac{\pi}{3\sqrt{2}}$ and therefore the packing density is 0.74.

The atoms occupy 74% of the volume, with the remaining 26% being empty space between atoms.

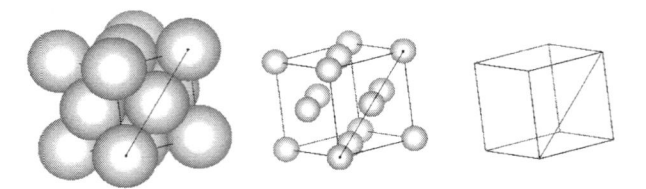

Figure 2.21. *Representations of a cubic F system*

The same reasoning can be applied to calculate the packing density for the cubic I/body-centered cubic system ($\pi\frac{\sqrt{3}}{8}$ or 0.68) or for the hexagonal close-packed system ($\frac{\pi}{3\sqrt{2}}$ or 0.74).

NOTE 2.6.– It is obvious that packing density is always <1 and it can be memorized in a first approximation as $\frac{3}{4}$, $\frac{2}{3}$ and $\frac{1}{2}$, values which are close to 0.74 for the hexagonal close-packed system, to 0.68 for body-centered cubic system and to 0.52 for non-close-packed system, respectively.

2.2.5. Designation of planes in a crystal

A plane can be completely defined by three arbitrary non-collinear points in the respective plane, given the intersections with the axes of coordinates, for example 1, 2 and 3. Rather than locating the plane directly by the coordinates of its points of intersection with the axes (1, 2 and 3), Miller indices are used. These indices facilitate the study of unit cells, as they allow for the definition of a plane that is parallel to the designated plane, which belongs to the unit cell (see Figure 2.22).

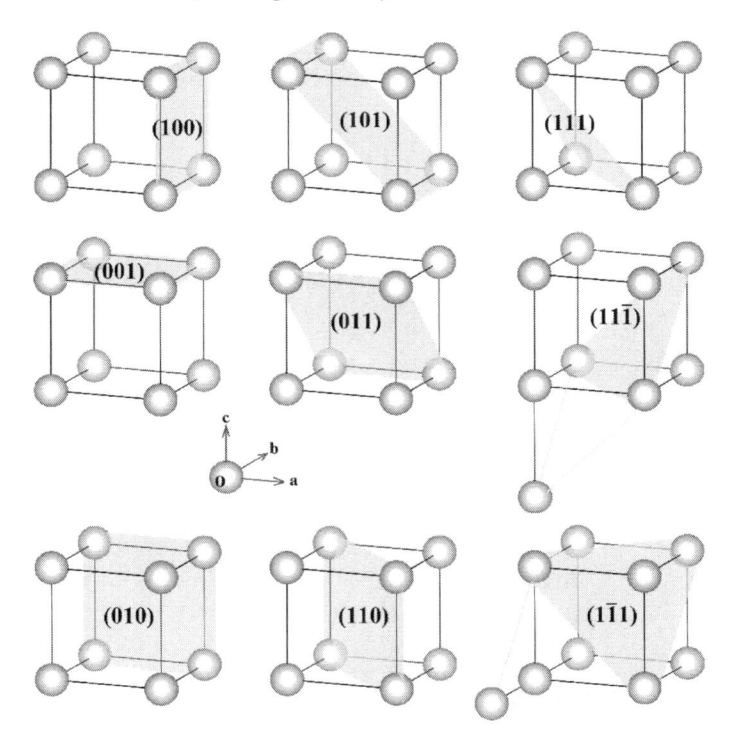

Figure 2.22. *Representations of planes in a cubic P system*

The three Miller indices of this plane are obtained as follows:

– The Miller indices of a family of crystallographic planes (with reference to a lattice) are integers that are in the same ratio as the reciprocal of the lengths of the planar intercept for the axes (namely $\frac{1}{1}$, $\frac{1}{2}$ and $\frac{1}{3}$) for an arbitrary plane that does not contain the origin.

– Integers that are in the same proportions as the three previous fractions should be determined, which amounts to bringing them to a common denominator (namely $\frac{6}{6}$, $\frac{3}{6}$ and $\frac{2}{6}$), which leads to 6, 3 and 2, and these are the Miller indices of the plane denoted by these figures enclosed within parentheses, not separated by spaces (632).

– It should be noted that if the plane is parallel to one of the axes, its intercept may be considered infinite; therefore, the corresponding index is null.

– If the plane intersects the axis on the negative side, the corresponding index is negative and in this case, the minus sign can be placed before the index (h–kl) although it is commonly placed over it (h$\bar{k}$l).

The crystal being infinite, for any plane there is an infinite number of parallel planes.

The distance between two closest parallel planes is generally useful and it is called interplanar distance denoted by d_{hkl}.

The higher the Miller indices, the closer the planes (and the smaller d_{hkl}) and for cubic packings, the following relation is applicable:

$$d_{hkl} = \frac{a}{\sqrt{h^2+k^2+l^2}} \qquad [2.4]$$

Using this relation, it can be verified that interplanar distances for (100), (010) and (001) planes are identical. Moreover, if (hkl) planes are considered, all the planes resulting from index permutation or index sign change have identical interplanar distances {(hkl), (hlk), (lkh), etc.}.

A family of planes denoted by {110} comprises the planes (110), (10$\bar{1}$), (0$\bar{1}\bar{1}$), ($\bar{1}$01), etc.

Rapid identification of planes in a unit cell is essential and the representations of various planes of a diamond cubic crystal in Figure 2.23 are helpful in getting acquainted with this notation.

In Figure 2.23, the drawing represents the planes from left to right (400), (200) and (100) in the first unit cell (1) and the planes (211), (121) and (112) for the next unit cells (2), (3) and (4), respectively.

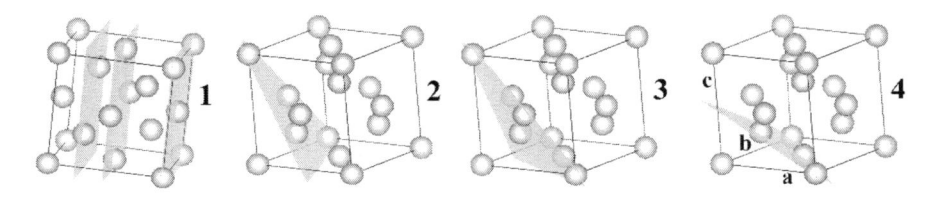

Figure 2.23. *Representations of planes in a diamond cubic system*

Given that the plane nomenclature will be frequently used in the following descriptions, its memorization is recommended.

Directions designating a vector perpendicular to a plane are also used, for example, the direction given by [320] is the line that passes through the point of coordinates 3, 2 and 0 and therefore is perpendicular to the (320) plane.

While this representation by the same figures may seem surprising, the definitions lead to identifying a plane and its perpendicular irrespective of the plane defined by these Miller indices (one more property of these indices).

> NOTE 2.7.– For example, a plane P (containing the three gray atoms) intersects the axes at coordinates 1 for a, 1.5 for b and it is parallel to c, therefore its coordinate is ∞. The reciprocals of these numbers are 1/1, 1/1.5 and 1/∞; therefore, the three proportional integers are (after having brought them to common denominator) 3/6, 2/6 and 0/6, which yield 3, 2 and 0, and the plane is denoted as (320).

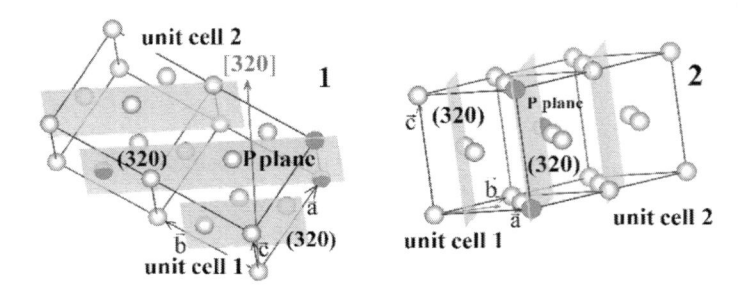

Let us consider two cubic F unit cells and the plane (320) of the two unit cells. It can be noted that the plane (P) of the second unit cell crosses the first one. Similarly, the plane of the third unit cell crosses the second one. Many planes may cross one unit cell without belonging to it and they are denoted starting from a neighboring unit cell.

2.2.6. Surface density

Packing density or volume density is proportional to the space filled by the atoms in a unit cell. The question arises if for a given unit cell density is the same along the chosen planes.

It is simple to calculate this surface density, as it amounts to calculating the surface area of n circles present in a given surface:

$$n \pi r^2 \qquad [2.5]$$

This value should be divided by that of the area of the considered surface, which is generally square, rectangle or triangle.

The densest planes have generally low indices, as the atoms have to be as close as possible (the higher the indices of a plane, the farther away from that plane are the atoms); therefore, they can be easily obtained if the (100), (110) and (111) planes are considered.

2.2.6.1. Surface density of the cubic P unit cell

If a cubic P system (2) is observed (see Figure 2.24), along the (100) plane represented in (1), it can be noted that all the atoms are in contact, while along the (110) plane represented in (3), the atoms are in contact vertically by twos. The (111) plane only contains three distant atoms (5 and 6) and its density is not high.

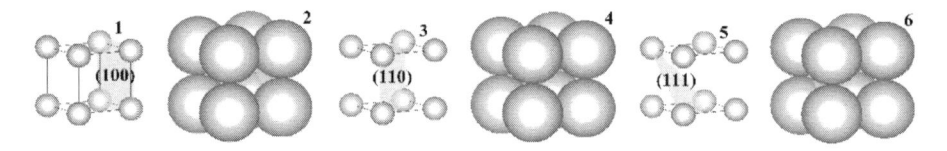

Figure 2.24. *Representations of dense planes in a cubic P system*

Knowing that a is equal to 2r and there is only one atom per plane (one-quarter of each of the four atoms is in the rectangle), surface density is given by the calculation:

– For the (100) plane $1\,\pi\dfrac{r^2}{a^2} = \dfrac{\pi.r^2}{(2r)^2} = \dfrac{\pi}{4}$ or 0.79

– For the (110) plane $1\,\pi\dfrac{r^2}{a.a\sqrt{2}} = 1\,\pi\dfrac{r^2}{(2r)^2\sqrt{2}} = \dfrac{\pi}{4\sqrt{2}}$ or 0.56.

A surface density of 0.79 is obtained for the (100) planes, which are high-density planes that are found in cubic P crystals (see Figure 2.28).

2.2.6.2. *Surface density of the cubic I unit cell*

When observing (see Figure 2.25) a cubic I system (1) along a (100) plane, it can be noted that atoms are not in contact (2) while along the (110) plane all the atoms are in contact with the central atom (3 and 4).

As previously, the (111) plane only contains three distant atoms and does not have high density (5 and 6).

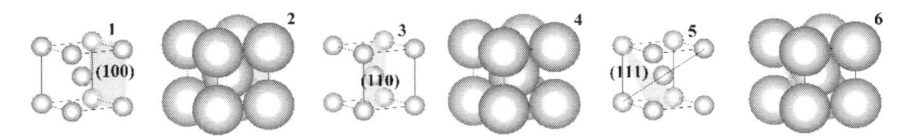

Figure 2.25. *Representations of dense planes in a cubic I system*

Knowing that the atoms are in contact along the diagonal (5) of the cube, the length of the diagonal $a\sqrt{3}$ can be written as 4r and the surface density results from the following calculation:

– For the (100) plane: $1\,\pi\,\dfrac{r^2}{a^2} = 1\,\pi\,\dfrac{r^2}{\left(\frac{4r}{\sqrt{3}}\right)^2} = \dfrac{3\pi}{16}$ or 0.59;

– For the (110) plane: $2\,\pi\,\dfrac{r^2}{a.a\sqrt{2}} = 2\,\pi\,\dfrac{r^2}{\left(\frac{4r}{\sqrt{3}}\right)^2\sqrt{2}} = \dfrac{\pi}{2\sqrt{2}}$ or 0.83.

A surface density of 0.59 results along the (100) plane but this is not the highest possible value, as for the (110) planes, the value 0.83 has been obtained.

These high-density planes are therefore visualized in cubic I/body-centered cubic crystals (see Figure 2.28).

2.2.6.3. Surface density of the cubic F unit cell

When observing (see Figure 2.26) a cubic F system (1) in the (100) plane, it can be noted that the four atoms are in contact with the central atom (2), whereas in the (111) plane, each atom is in contact (3 and 4) with the six atoms around it.

The (110) plane is not high in density, as only the horizontally aligned atoms are in contact.

Knowing that the atoms are in contact along the diagonal of one face of the cube ($a\sqrt{2}$) and that along this diagonal four radii are cumulated, this leads to $a = 2r\sqrt{2}$, and the surface density is given by the following calculation:

– For the (100) plane: $2\,\pi\,\dfrac{r^2}{a^2}$ or $2\,\pi\,\dfrac{r^2}{(2r\sqrt{2})^2}$ or $\dfrac{\pi}{4}$ or 0.79;

– For the (111) plane: $2\pi\,\dfrac{r^2}{\frac{1}{2}\left(a\sqrt{2}.a\sqrt{2}\frac{\sqrt{3}}{2}\right)}$ or $2\pi\,\dfrac{r^2}{\frac{1}{2}\left(2(r\sqrt{2})^2\frac{2\sqrt{3}}{2}\right)}$ or $\dfrac{\pi}{2\sqrt{3}}$ or 0.91.

A surface density of 0.91 results for the (111) planes, which are high-density planes found in cubic F crystals (see Figure 2.27).

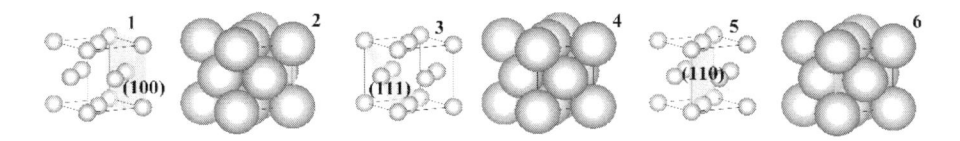

Figure 2.26. *Representations of dense planes in a cubic F system*

2.2.6.4. *Crystal faces*

Crystal faces are often formed along the highest-density planes and the cubic F system can yield tetrahedrons or octahedrons due to the high density (0.91) of the (110) planes. But it can also yield rhomboidal dodecahedrons due to the density (0.56) of the (111) planes.

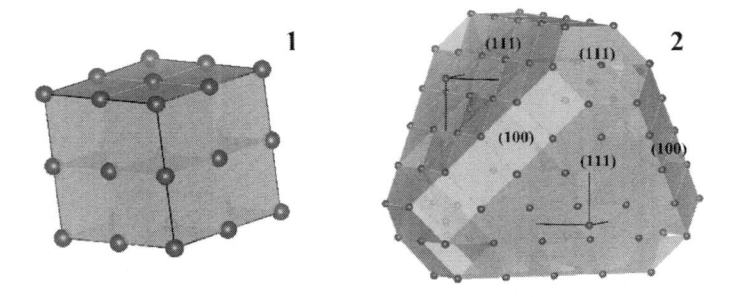

Figure 2.27. *Fictitious polyhedrons obtained from a cubic P system (1) and a cubic F system (2). For a color version of this figure, see www.iste.co.uk/valls/inorganic.zip*

Two examples of fictitious polyhedrons, counting the least possible number of atoms and obtained by combining various planes of a cubic lattice, are proposed in Figure 2.27, where the cube corresponding to the unit cell is highlighted (in black).

The first lattice (1) can be considered cubic P ((100) planes on all the faces) and the second (2) can be considered cubic F ((111) planes on the majority of faces) and a truncated tetrahedron can be spotted on the edges by the (100) planes and on the vertices by the (111) planes.

The rhomboidal dodecahedron can be obtained from the (110) planes and it leads to 12 faces, all rhomboid-shaped, which may lead to the idea of a cubic I lattice. Indeed, a cubic I unit cell can be represented at the bottom of

(1) and it can be noted that the four unit cells are sufficient to highlight this dodecahedron (see Figure 2.28).

The second representation shows a truncated cube; this image has been obtained starting from the (100) planes and is derived from a cubic P system of (100) planes on the majority of faces, with the vertices being truncated by (111) planes and the edges by (110) planes.

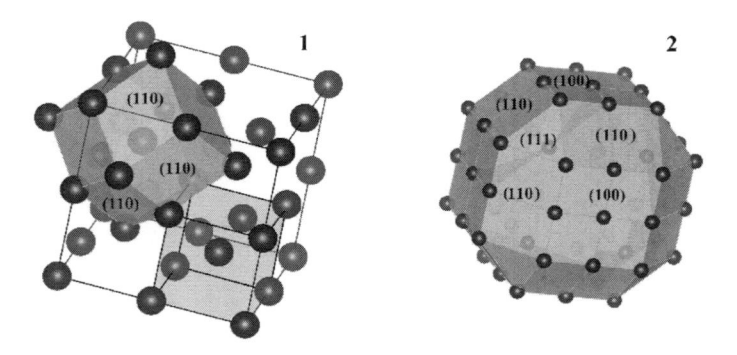

Figure 2.28. *Fictitious polyhedrons obtained from a cubic I system (1) and from a cubic P system (2). For a color version of this figure, see www.iste.co.uk/valls/inorganic.zip*

NOTE 2.8.– Innumerable combinations of these high-density planes can be imagined, leading to crystals that are present in nature under most varied forms. Four regular forms are proposed in the figure below.

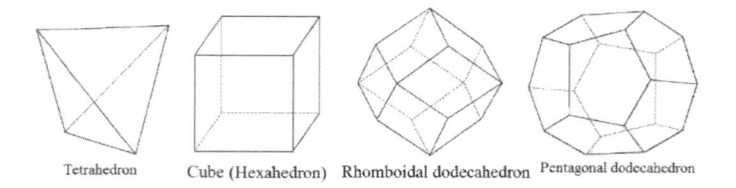

Tetrahedron   Cube (Hexahedron)   Rhomboidal dodecahedron   Pentagonal dodecahedron

The simplest, though not always the most common, are the tetrahedron and the cube for the cubic system.

Other forms are surprising, such as the already described pentagonal dodecahedron, which has faces with five edges!

2.3. Representation of metallic crystals

The representation of packings requires normalization, to allow for a widely comprehensible description. Various planes of a crystal are denoted according to the previously defined Miller indices and the name of the unit cell is assigned to a simple geometric volume associated with a crystal.

The unit cell is therefore a box with six faces closely associated with translation, since a crystal is constituted from an assembly of patterns that are periodically repeated in the three directions of space.

2.3.1. Definition of the unit cell

A crystal is a solid with regular and periodic structure, comprising a significant number of atoms, molecules or ions that are orderly packed. This lattice can be divided into identical volumes that are infinitely repeated, which are the unit cells.

The unit cell should be conceived of as an elementary box with six parallel faces, two to two, (rectangular parallelepiped or not) with equivalent nodes at each vertex. The unit cell corresponding to a packing has the least possible volume as well as the highest number of symmetries of this packing, the edges being a, b and c (see Figure 2.29) and the angles α, β and γ (each angle is opposite to an edge).

A given lattice has a high number of possible parallelepiped unit cells; therefore, the conventional unit cell is used. This is the simplest possible unit cell that still accounts for the orientation symmetry of the structure in the best possible way. If several unit cells satisfy these criteria, preference will be given to those whose angles are closest to 90°.

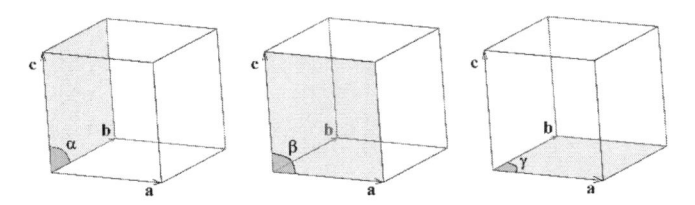

Figure 2.29. *Representation of a cubic unit cell with its parameters*

A simple unit cell contains a formular group, a multiple unit cell contains several formular groups; for example, NaCl contains four Na^+ cations and four Cl^- anions and therefore four formular groups. The number of formular groups is called the multiplicity of the unit cell and is denoted by the letter Z.

In a more mathematical approach, the unit cell is defined by three vectors $\vec{a}, \vec{b}$ and $\vec{c}$ that are repeated by translation along a regular lattice in order to form the crystal (see Figure 2.30). The unit cells are connected by translations and rotations (of angle 360°/n) but because of crystal periodicity, the values of the order n of rotation are limited to 1, 2, 3, 4 and 6 (contrary to molecules, for which it can take an arbitrary value).

2.3.1.1. *Bravais lattices*

The simplest approach to describing a crystal uses Bravais lattices, which are regular distributions of points called lattice nodes.

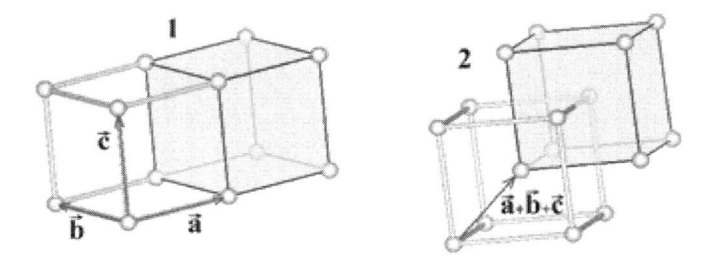

Figure 2.30. *Representation of lattice translations of cubic P and cubic I systems*

These points result from the translations of vectors $\vec{a}, \vec{b}$ or $\vec{c}$ (arbitrary edge of the cube) for cubic P structure or of vector ½ $\vec{a}$ + ½ $\vec{b}$ + ½ $\vec{c}$ (half-diagonal of the cube) for the cubic I system (see Figure 2.30).

Bravais proposed seven types of lattices, from the most symmetrical one, the cubic structure, to the least symmetrical one, the triclinic structure. However, one lattice can be declined in four modes and if the starting point is a cube, for example, this can be primitive (1), centered (2) or all-face-centered (3), as shown in Figure 2.31.

The two-face-centered cubic system (4) is equivalent to quadratic P. Therefore, symmetry leads to equivalent lattices and the variety of seven lattices unfolded in four modes leads to only 14 Bravais lattices instead of the expected 28, since some lattices are equivalent.

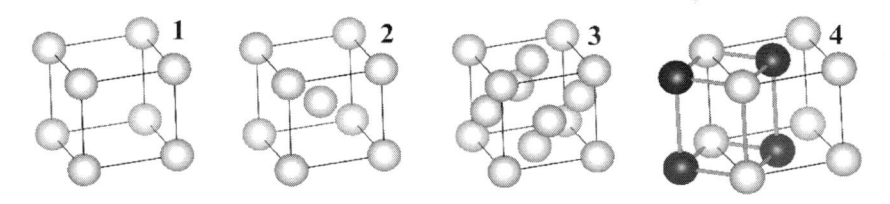

Figure 2.31. *Representation of cubic lattices modes*

The two-face-centered mode is denoted by the letter C, as centered faces are generally C faces. However, for the orthorhombic lattice, there can be A or B modes if the centered faces are A or B.

All systems have a primitive form P but not necessarily all the other derived forms. The cubic system covers three types of lattices: P, I and F; the tetragonal or quadratic system has P and I lattices; the orthorhombic system has four lattices: P, I, F and C (or A or B); the hexagonal system has one lattice P; the monoclinic system has two lattices: P and C (or A or B); the trigonal P and R (rhombohedral with two equal edges and angles of the unit cell) has two lattices and the triclinic has only one P lattice.

2.3.1.2. *Space groups and point groups*

Bravais lattices (BL) allow for a rapid description of the simplest packings but they are not sufficient for describing all crystalline structures. Indeed, the latter can comprise patterns containing groups of atoms and not simply atoms, at each lattice node.

The space group (SG) of a crystalline system covers all the symmetry operations of the point group or crystalline class (CC) and the translation operations.

The crystalline class (CC) of symmetry of a crystalline system is the group, in the mathematical sense, covering all the symmetry operations with respect to which a lattice node is invariant. This node is therefore situated at the intersection of all symmetry operations, translation not included.

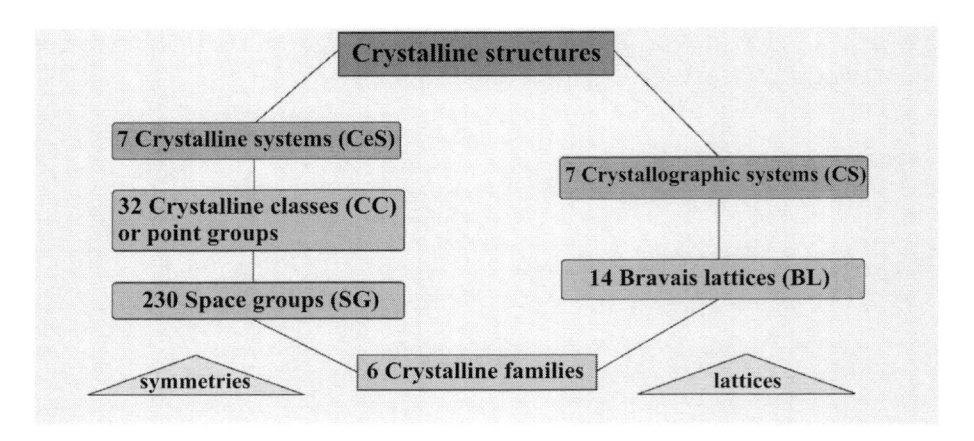

Figure 2.32. *Crystalline systems, classes and space groups*

It can be shown that only certain combinations of point symmetry operators (rotation axes, symmetry plane, inversion, etc.) are possible. They lead to 32 combinations that are the 32 crystalline classes for which symmetry is compatible with the lattice (see Figure 2.32).

By 1890, Fedorov and Schoenflies had independently proved the existence of these 230 space groups, which cover all the possible combinations of lattices and symmetry operations.

Given their significant number, space groups can be subdivided into various categories:

– according to a symmetry-based reasoning, there are 32 crystalline classes (CC), the only ones compatible with the three-dimensional periodicity, out of which seven are holohedrons and allow for the definition of seven crystalline systems (CeS), whereas the others are qualified as merohedral;

– according to a Bravais lattice (BL)-based reasoning, there are 14, which obviously yield seven crystallographic systems (CS);

– the seven systems are drawn from six families, as the hexagonal system can be considered a specific case of the trigonal system. The trigonal system is a hexagonal system that has lost the axis of the sixth order in favor of an axis of third order.

Both approaches are needed; if lattices are considered with patterns, the 14 lattices are insufficient (symmetry is generally reduced by these patterns) and 230 space groups are obtained as a result of group theory (microscopic symmetry is the term employed).

These 230 space groups include 32 crystalline classes characterized by macroscopic symmetry and seven crystalline systems characterized by elementary or primitive unit cells. The lattice-based approach, which is more intuitive, will be preferentially used, and it is sufficient for simple crystals.

There are seven crystalline systems (CeS) or holohedral crystalline classes (CC) or crystallographic systems (CS) having all lattice symmetries for the following types of system:

– cubic: the holohedral class is denoted as m-3m and has three axes of the fourth order that are perpendicular to the faces of the cube (passing through their center) and four axes of third order that are the diagonals of the cube;

– tetragonal (or quadratic): the holohedral class is denoted as 4/mmm and presents an axis of the fourth order parallel to c, the (100) planes are main mirrors and the (110) planes are secondary mirrors;

– orthorhombic: the holohedral class is denoted as mmm and has axes of the second order perpendicular to one another;

– hexagonal: the holohedral class is denoted as 6/mmm and has an axis of the sixth order parallel to the c axis, the (100) planes are main mirrors and the (110) planes are secondary mirrors;

– monoclinic: the holohedral class is denoted as 2/m and has an axis of the second order along the c (or b) axis;

– rhombohedral: the holohedral class is denoted as −3m and has an axis of the third order (it is the c axis of the hexagon with a pattern that limits the order of the main axis to 3);

– triclinic: the holohedral class is denoted as −1.

Table 2.1 summarizes seven systems, 14 lattices and 32 crystalline classes (CC) as well as the space groups (SG) and their relative positioning. The axes and their direction have been indicated, as well as the number of space groups for each point group.

7 systems	14 lattices	32 PG	Axes	SG
Cubic	cP, cI, cF	23, m3, 432, −43m, m-3m	2 or −2 ; 4 or −4 following [100] [010] or [001] and 4 directions 3 or −3	36
Tetragonal (quadratic)	tP, tI	4, −4, 422, 4mm, −42m, 4/m, **4/mmm**	4 or −4 following [001] then 2 or −2	68
Orthorhombic	oP, oI, oF, oC	222, mm2, **mmm**	3 directions ⊥ 2 or −2	59
Monoclinic	mP, mC	2, m, **2/m**	2 or −2 following [010] or [001]	13
Hexagonal	hP	6, −6, 6/m, 622, 6mm, −62m, **6/mmm**	6 or −6 following [001] then 2 or −2	27
Rhombohedral	hR	3, 32 ,3m,−3, **−3m**	3 or −3 following [001] then 2 or −2	25
Triclinic	aP	1, −1	all directions	2

Table 2.1. *Systems, lattices, point groups and space groups*

This inventory can be summarized by combining the 32 classes and four modes:

– For the triclinic and hexagonal only of P type, there are two and eight CC;

– For the trigonal, which is only of P and R type, there are 13 CC;

– For the monoclinic, P and C are possible and double the CC to 6;

– For the orthorhombic, of P, C, I and F types, there are 12 CC;

– For the tetragonal, there are 16 CC and 15 for the cubic system.

System	Class	2	3	4	6	Planes	Center	Crystals
Cubic	m-3m	6	4	3	-	9	yes	P-ReO$_3$, P-CsCl, F-NaCl, F-CaF$_2$, F-Li$_2$O
	-43m	3	4	-	-	6	-	F-ZnS
Tetragonal	P4/mmm	4	-	1	-	5	yes	P4$_2$/mnm TiO$_2$
Hexagonal	P6/mmm	6	-	-	1	7	yes	P6$_3$/mcm NiAs
	6mm	-	-	-	3	6	-	P6$_3$mc ZnS, P6$_3$mc NiAs
Trigonal	-3m	3	1	1	6	3	yes	R-3m CdCl$_2$, P-3m$_1$ CdI$_2$

Table 2.2. *Crystalline characteristics of typical crystals*

Overall, there are 72 point groups, a number that is far from 230; therefore, translations are needed from triclinic to cubic (0, 4, 37, 20, 6, 4, 14 or 85 in total) then rotation-gliding, respectively (0, 3, 10, 32, 6, 15, 7 or 73 in total). This yields 72 + 73 + 85, which amounts to 230 space groups.

Only four classes, six point groups and 10 space groups will be used to describe the typical structures presented in this work; therefore, full knowledge of group theory is not essential (see Table 2.2).

This description contributes nevertheless to an awareness of notions of crystallography which, although purely mathematical, are required for understanding crystals.

2.3.1.3. *Description of the codified notation*

It is sometimes difficult to understand the subtleties associated with the symbolism of the crystal classification into space groups. The objective here is to describe it as effectively as possible within the limits of subsequent needs.

The lattice symbol is preceded by letters P, I, C, F or R, which are associated with the following modes: primitive (P), body-centered (I), two-face-centered (C or A or B), all-face-centered (F) or rhombohedral (R).

Direction	First	Second	Third
cubic	[100], [010], [001]	[111], [1$\bar{1}$1], [$\bar{1}$11], [$\bar{1}\bar{1}$1]	[110], [1$\bar{1}$0], [101], [10$\bar{1}$], [011], [01$\bar{1}$]
tetragonal	[001]	[100], [010]	[110], [1$\bar{1}$0]
orthorhombic	[100]	[010]	[001]
hexagonal	[001]	[100], [010], [110]	[210], [120], [1$\bar{1}$0]
monoclinic	[010]	-	-
trigonal (P)	[001]	[100], [010], [110]	-
trigonal (R)	[111]	[1$\bar{1}$0], [01$\bar{1}$], [10$\bar{1}$]	-

Table 2.3. *Conventional order of symmetry directions*

The lattice notation uses two or three symbols corresponding to the symmetry elements in the characteristic directions. In international notation, Hermann–Mauguin symbols indicate the elements of the symmetry operations of a space group along each symmetry direction (characteristic of each crystallographic system).

Hermann–Mauguin symbols are oriented, and the orientation of each symmetry element can be read from the symbol, knowing that symmetry directions are given in a standard order (see Table 2.3) in each crystallographic system. Since it presents no direction, the triclinic system is not shown.

For non-cubic systems, a first symbol denotes the order of the main axis, if it is the only one and equal to:

– 1 or −1, for triclinic system;

– 2 or m once, for monoclinic system;

– 3 or −3, for trigonal system;

– 4 or −4, for tetragonal system;

– 6 or −6 for hexagonal system.

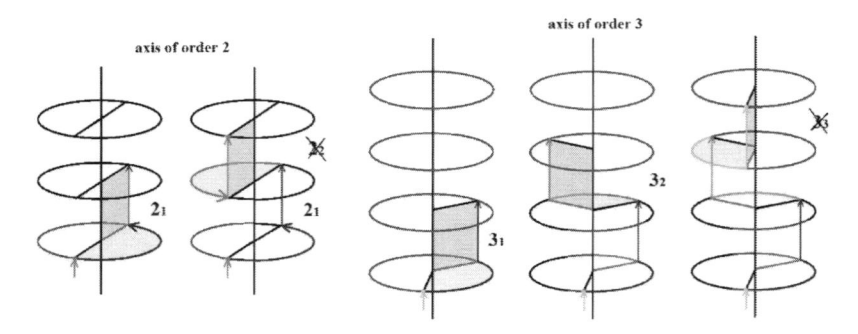

Figure 2.33. *Axes of rotation-glide of orders 2 and 3*

The axis order may feature an index when it is associated with a rotation of order n (see Figure 2.33), which is followed by a glide of ½ for order

2 (the only coded possibility 2_1 or 21). On the contrary, for order 3, there may be a glide of 1/3 (coded 3_1 or 31) or 2/3 (coded 3_2 or 32), the third possibility being a return to the initial point (coded 3_3); therefore, there are only two possibilities. Similar reasoning can be applied to orders 4 and 6, which give three and five possible glides.

The second symbol denotes the axis of symmetry in a plane that is perpendicular to the main axis (2 or m), except for the cubic system for which the second symbol is always 3 or −3.

When applicable, the third symbol characterizes an axis whose existence ensues from the first two, and is 2 or m.

Cubic systems are assigned specific nomenclature, featuring two or three symbols (see Table 2.3) in order to avoid confusions with symbols that have already been used:

– the first symbol characterizes the three directions of type (100);

– the second symbol is always 3 or −3 (four directions of type (111));

– the third symbol characterizes the six directions of type (110) (see Table 2.3).

This classification allows for rapid identification of a cubic system, the only one featuring four ternary axes indicated by the second figure. Similarly, the hexagonal system is the only one having a hexary axis (explaining the number 6) or the trigonal, the only one starting by number 3. The tetragonal starts by number 4 and, finally, the orthorhombic and the monoclinic start by 2 or m but they count three and two successive symbols, respectively.

It can be noted that if the mirror is perpendicular to the axis of order n, it is denoted n/m, if the mirror passes through the axis of order n, it is denoted nm and if the translation refers to the three characteristic directions, they are assigned an index.

A reflection plane inside the point groups can be replaced by a glide (or translation) plane, denoted by a, b or c along half a vector of the unit cell (see Table 2.4).

Symbol	Symmetry	Glide	Orientation	
E	Double glide	½ period	2 axes ⊥	O
a		a/2	(010) or (001); (01$\bar{1}$)	
b		b/2	(100) or (001); ($\bar{1}$01)	M, Q, O, C; R
c	Axial glide	c/2	(100) or (010); (1$\bar{1}$0)	
C		c/2	(1$\bar{1}$00) or (11$\bar{2}$0)	H
			(0$\bar{1}$0) (110)	Q
N	Diagonal glide	(b + c)/2	(100)	
		(a + c)/2	(010)	M, Q, O, C
		(a + b)/2	(001)	
D	Diamond glide	(b ± c)/4	(100)	
		(a ± c)/4	(010)	O, C
		(a ± b)/4	(001)	
		(± a + b ± c)/4	(01$\bar{1}$) (011)	C
		(± a ± b + c)/4	($\bar{1}$01) (101)	
		(a ± b ± c)/4	(1$\bar{1}$0) or (110)	C, Q

Table 2.4. *Symbols of glide by order of priority*

There is also an n-glide, which takes place along half of the diagonal of one face of the unit cell and the d-glide, which is along a quarter of a diagonal of one face or of the cube (this glide appears in the diamond structure).

Finally, there is the e-glide (which refers exclusively to the orthorhombic C) and one translation of half a period along two perpendicular directions.

It should be noted that d mirrors exist only for orthorhombic F, tetragonal (or quadratic) I and cubic F and I lattices.

The corresponding point group results from replacing the glide symmetry elements by the corresponding point symmetry element (m). For example, the orthorhombic P na2$_1$ is simply a variety of the P mm2 point group.

A summary of space groups is shown in Figure 2.34, where colors contribute to a better visualization of their characteristics.

Knowledge of Metallic Crystals 91

system	32 PG	230 EG	groups	mode					
triclinic system	1	1	proper	P1					
	-1	2	centrosymmetric	P-1					
monoclinic system	2	3-5	proper	P2 C2	P2₁				
	m =-2	6-9	impropers	Pm Cm	Pc Cc				
	2/m	10-15	centrosymmetric	P2/m C2/m	P2/c C2/c	P2₁/m	P2₁/c		
orthorhombic system	222	16-24	proper	P222 C222 I222 F222	P222₁ C222₁ I2₁2₁2₁	P2₁2₁2	P2₁2₁2₁		
	mm2	25-46	impropers	Pmm2 Cmm2 Imm2 Fmm2	Pcc2 Pmc2₁ Ccc2 Cmc2₁ Iba2 Fdd2	Pma2 Pca2₁ Amm2 Ima2	Pnc2 Pmn2₁ Aem2	Pba2 Pna2₁ Ama2	Pnn2 Aea2
	mmm	47-74	centrosymmetric	Pmmm Cmmm Immm Fmmm	Pnnn Pmna Pnnm Cmcm Ibam Fddd	Pccm Pcca Pmmn Cmce Ibca	Pban Pbam Pbcn Cmcm Imma	Pmma Pccn Pbca Cccm Cmme	Pnna Pbcm Pnma Ccce
tetragonal system	4	75-80	proper	P4 I4	P4₁ I4₁	P4₂	P4₃		
	-4	81-82	impropers	P-4 I-4					
	4/m	83-88	centrosymmetric	P4/m I4/m	P4/n I4₁/a	P4₂/n	P4₂/m		
	422	89-98	proper	P422 I422	P42₁2 P4₁22 P4₁2₁2	P4₁22 P4₂2₁2	P4₂2₁2	P4₃22	P4₃2₁2
	4mm	99-110	impropers	P4mm I4mm	P4bm P4₂cm P4₂nm I4cm	P4cc P4₂bc I4₁md	P4nc I4₁cd	P4₂cm	P4₂nm
	-42m	111-122	impropers	P-42m P-42c I-42m	P-4m2 P-42₁m P-42₁c I-4c2	P-42c P-42₁c I-42d	P-4c2	P-4b2	P-4n2
	4/mmm	123-142	centrosymmetric	P4/mmm I4/mmm	P4/mmc P4/nmm P4₂/nnm I4/mcm	P4/nbm P4/ncc P4₂/mbc I4₁/amd	P4/nnc P4₂/mmc P4₂/mnm I4₁/acd	P4/mbm P4₂/mcm P4₂/nmc	P4/nnc P4₂/nbc P4₂/ncm
trigonal system	3	143-146	proper	P3 R3	P3₁	P3₂			
	-3	147-148	centrosymmetric	P-3 R-3					
	32	149-155	proper	P321 R32	P312	P3₁12	P3₁21	P3₂12	P3₂21
	3m	156-161	impropers	P3m1 P3m R3c	P31m R3c	P3c1	P31c		
	-3m	162-167	centrosymmetric	P-3m1 R-3m	P-3m1 R-3c	P-31c	P-3c1		
hexagonal system	6	168-173	proper	P6	P6₁	P6₂	P6₃	P6₄	P6₅
	-6	174	impropers	P-6					
	6/m	175-176	centrosymmetric	P6/m	P6₃/m				
	622	177-182	proper	P622	P6₁22	P6₂22	P6₃22	P6₄22	P6₅22
	6mm	183-186	impropers	P6mm	P6cc	P6₃cm	P6₃mc		
	-6m2	187-190	impropers	P-6m2	P-62m	P-6c2	P-62c		
	6/mmm	191-194	centrosymmetric	P6/mmm	P6/mcc	P6₃/mcm	P6₃/mmc		
cubic system	23	195-199	proper	P23 I23 F23	P2₁3 I2₁3				
	m-3 (2/m-3)	200-206	centrosymmetric	Pm-3 Im-3 Fm-3	Pn-3 Ia-3 Fd-3	Pa-3			
	432	207-214	proper	P432 I432 F432	P4₂32 I4₁32 F4₁32	P4₁32	P4₃32		
	-43m	215-220	impropers	P-43m I-43m F-43m	P-43n I-43d F-43c				
	m-3m	221-230	centrosymmetric	Pm-3m Im-3m Fm-3m	Pn-3n Ia-3d Fm-3c	Pm-3n	Pn-3m	Fd-3m	Fd-3c

Figure 2.34. *The 32 classes, 230 space groups and four modes (holohedral crystalline classes are bolded). For a color version of this figure, see www.iste.co.uk/valls/inorganic.zip*

NOTE 2.9.– Hermann–Mauguin symbols are often given in their abbreviated form and, when binary axes and mirrors coexist, it suffices to indicate the mirrors, as the axes are generated by combination, for example, 2/m2/m2/m yields mmm in abbreviated form.

The only exception is the 2/m group, which cannot be given in abbreviated form, as there is only one symmetry direction in the monoclinic crystallographic system.

2.3.1.4. *Description of the body-centered cubic unit cell*

The conventional body-centered cubic unit cell (1) contains two nodes, while the unit cell containing one node (2 and 3) is triclinic (a = b ≠ c and α = β = 54°7, γ = 90°), being therefore difficult to manipulate (see Figure 2.35). Adding one node (1) leads to obtaining the cubic symmetry, which is very easy to manipulate and describe.

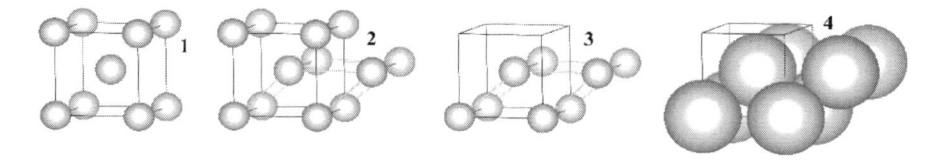

Figure 2.35. *Representation of the conventional elementary cubic I unit cell*

The cubic I system in which the central atom is in contact with the other eight atoms (1 and 2) and is located at the center of a cube should not be mistaken for the NN' packing, in which the central atom (3 and 4) is at the center of a tetragonal unit cell (see Figure 2.36).

This tetragonal unit cell (NN' packing) is not arbitrary and its $\frac{c}{a}$ ratio is $\sqrt{2}$, which is the condition for maximum contact between atoms.

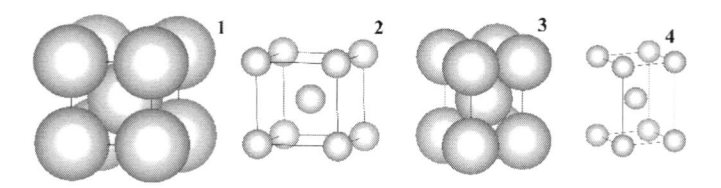

Figure 2.36. *Representation of conventional cubic I and tetragonal unit cells*

2.3.1.5. Description of all-face-centered cubic unit cell

The conventional all-face-centered cubic unit cell (1 and 2) (see Figure 2.37) contains four nodes (see section 2.2.3). The elementary unit cell containing one node is rhombohedral (3 and 4) and visible in the cube representing the cubic F system. It is a specific rhombohedral system with an angle of 60° (3 and 4) but this volume is not the best when explaining the symmetry of the structure and it is not generally used.

In order to represent the cubic F unit cell (see Figure 2.38), since the atoms need to be in contact, a specific three-node rhombohedral unit cell (1) can be used.

By representing four atoms of the chosen unit cell (2) and 10 complementary atoms outside the unit cell (3), the cubic F unit cell is visualized by associating them (4). The levels of ABC packing (4) characteristic of cubic F have been highlighted. This could be imagined by considering the three atoms in contact of representation (2), forming an ABC packing.

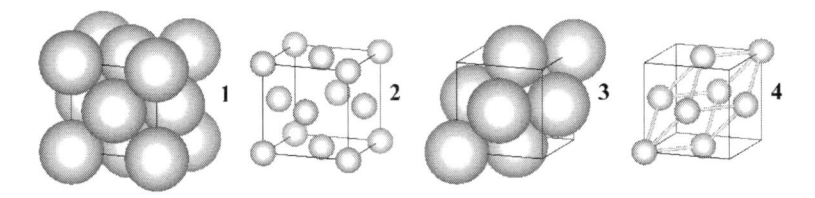

Figure 2.37. *Representation of conventional and elementary cubic F unit cells*

While these two representations of the cubic F unit cell are not generally used, they can serve to show that the choice of a unit cell is not simple and immediate and that various geometric figures can simultaneously be present in a packing.

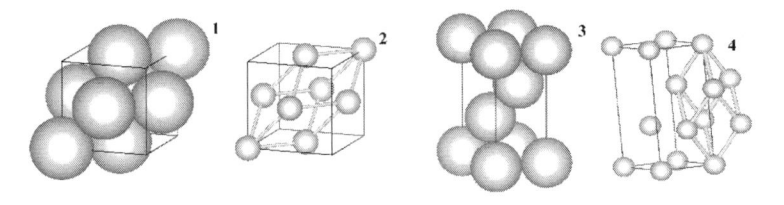

Figure 2.38. *Various representations of the cubic F unit cell*

Indeed, in a cubic F packing, a rhombohedron, a hexagon and a cube (with different multiplicity, obviously) can be found. The arbitrary rhombohedral unit cell does not necessarily ensure the contact of atoms along the vertical axis, although this may be achieved along horizontal planes (see Figure 2.39). A rhombohedral unit cell (1 and 2) has been represented, which allows for the visualization of a cubic F packing by representing (3 and 4) several atoms of neighboring unit cells.

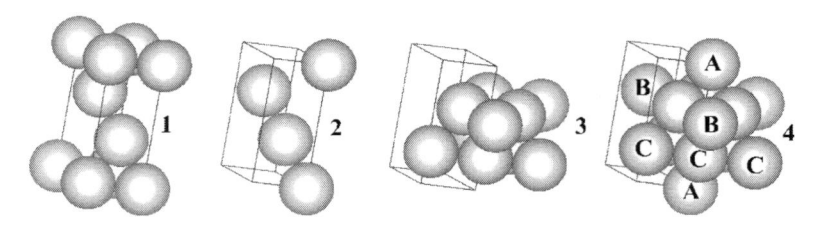

Figure 2.39. *Representation of the cubic F rhombohedral unit cell*

NOTE 2.10.– Can it be concluded that rhombohedral (1 and 2) and cubic F systems are identical? The answer is affirmative if the $\frac{c}{a}$ ratio is 2.449, as it is the value that ensures the contact of all the atoms of the rhombohedron.

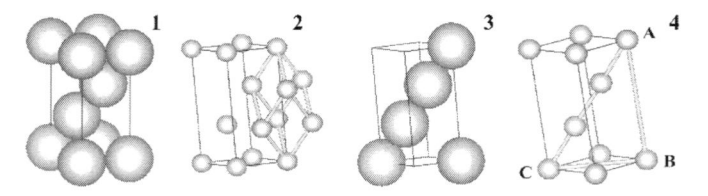

This figure (4) allows for the visualization of a right triangle ABC, in which based on the edge AC = 6r (3) and the edge BC = $2r\frac{\sqrt{3}}{2}$ (two times the height of the equilateral triangle formed by three atoms in contact), the length of the edge MN = $\sqrt{(6r)^2-(2r\sqrt{3})^2}$ = 4.899 is calculated for r = 1 and $\frac{c}{a} = \frac{c}{2r} = 2.449$.

The trigonal unit cell with a pattern formed of one atom is available in rhombohedral and hexagonal forms and, in Figure 2.40, a trigonal unit cell P (1) has been represented and several atoms of the lattice around it have been chosen, outside this unit cell, which facilitate the visualization of the hexagonal unit cell P (2).

Then, a trigonal unit cell R (3) with three atoms has been represented (which limits the order of the main axis to 3) around which several atoms of the lattice have been chosen, outside this unit cell, which facilitate the visualization of a cuboctahedron (4) that characterizes the ABC packing or the cubic F lattice (to the extent that the measure or $\frac{c}{a}$ is 2.449).

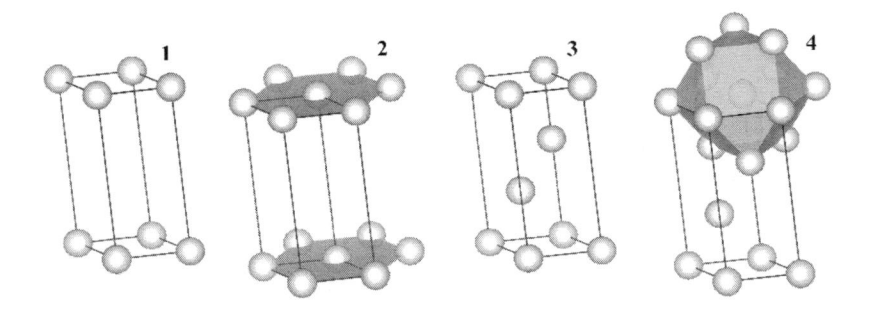

Figure 2.40. *Representation of trigonal P and R unit cell*

This language ambiguity related to trigonal, rhombohedric and hexagonal systems is often manifest but ambiguity is lifted as soon as the pattern involves several atoms (see Figure 2.41).

It is difficult to visualize the pattern in the conventional unit cell (1) but if several atoms are chosen outside the unit cell (2), this three-atom pattern is visualized. This reduces the hexary axis (3) into an axis of order 3 and the trigonal (1) is differentiated from the hexagonal (4), which has a hexary axis represented in blue (4).

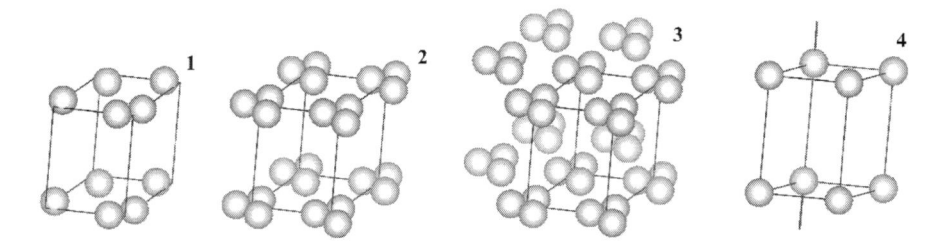

Figure 2.41. *Representation of a trigonal unit cell with a three-atom pattern*

The variety of systems renders their differentiation complex, as a triclinic system to which is added the same value to the three edges becomes rhombohedral. Furthermore, if angles are fixed at 90°, it becomes cubic P, whereas if the angle is fixed at 60°, it becomes cubic F.

On the basis of an axis of order 6 with an angle of 120° and the others at 90°, the hexagonal system is obtained; however, if the pattern decreases the order of the axis to 3, then it becomes trigonal. Moreover, if the ratio $\frac{a}{c} = 2.449$, it becomes cubic F.

2.3.2. Geometry of simple polyhedrons

Understanding unit cells requires good visualization of simple volumes as well as their properties. It is therefore useful to memorize certain properties of the cube, tetrahedron and hexagon.

2.3.2.1. The cube

The cube has six faces, eight vertices and 12 edges, which have to be mentally visualized in a simple and effective manner.

The cube is memorized by the two top and bottom faces that are necessarily connected by four lateral faces or 2 + 4 or six faces. There are four vertices on the top face and four vertices on the face below, in total 4 + 4 or eight vertices. There are four edges on the top face and four vertices on the face below, connected by four lateral edges, in total 4 + 4 + 4 or 12 edges.

The cube presents four axes of symmetry of order 3, three axes of order 4 and six axes of order 2, as shown in Figure 2.42.

NOTE 2.11.– It should be noted that the axes of symmetry of the cube are four of order 3 (1), three of order 4 (2) and six of order 2 (3) and therefore the product in each case amounts to 12, which is also the number of edges (but this is a coincidence of figures).

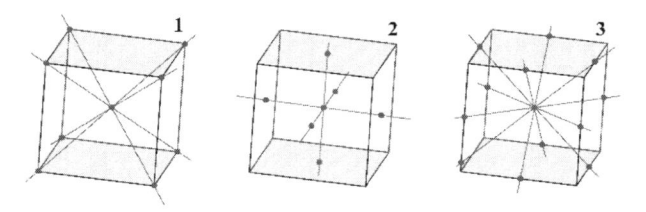

Figure 2.42. *Representation of the axes of symmetry 2, 3 and 4 of the cubic unit cell*

The diagonal of the face is equal to $a\sqrt{2}$ (diagonal of a square) and the diagonal of the cube is equal to $a\sqrt{3}$ (see Figure 2.43). To verify these results, it is sufficient to apply the Pythagoras theorem to the lines in the figure, since the square of the diagonal of the cube is the sum of the squares of a and $a\sqrt{2}$, which can be written as

$$a^2 + \left(a\sqrt{2}\right)^2 \quad \text{or} \quad a^2 + 2a^2 \quad \text{or} \quad 3a^2 \qquad [2.6]$$

whose square root yields the length of the diagonal:

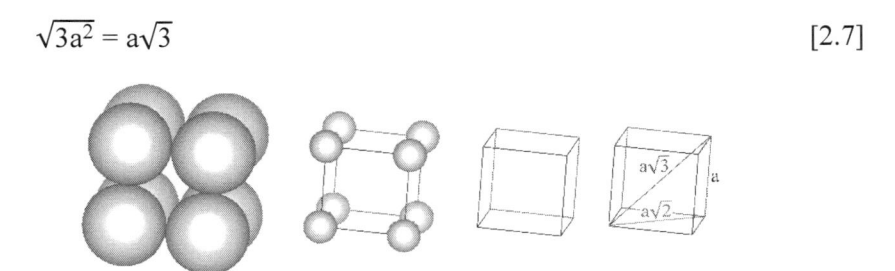

$$\sqrt{3a^2} = a\sqrt{3} \qquad [2.7]$$

Figure 2.43. *Representation of the cubic P unit cell and its diagonals*

It is worth noting that the diagonal of the face of the cube is equal to $a\sqrt{2}$ and that the diagonal of the cube is $a\sqrt{3}$.

2.3.2.2. The tetrahedron

The tetrahedron has four faces, four vertices and six edges.

The tetrahedron (2 and 5) can be memorized by an MNP face (1) in figure 2.44, which is the base of an (MNP) pyramid, which has by necessity three lateral faces or 1 + 3, which means four faces.

There are three vertices on the base and one above, or 3 + 1 or four vertices (2).

There are three edges on the base (1), connected by three lateral edges, or 3 + 3 or six edges.

Since the reasoning applied to the cube is easy, the tetrahedron can be placed in a virtual cube (3 and 5).

The relations valid for a cube can be applied, knowing that the only known data of this virtual cube is the diagonal of a face (3), which equals the double radius of the atom (2r). If the edge of this virtual cube is a', then:

$$a'\sqrt{2} = 2r \quad \text{or} \quad a' = 2\sqrt{2} \qquad [2.8]$$

It is then simple to calculate the distance between an atom and the center of a tetrahedron (which is also the center of the virtual cube) as it is the half-diagonal of this cube:

$$\text{or } a'\frac{\sqrt{3}}{3} \quad \text{or} \quad r\sqrt{2}\frac{\sqrt{3}}{2} \quad \text{or} \quad r\frac{\sqrt{3}}{\sqrt{2}} \qquad [2.9]$$

The height of the tetrahedron is calculated after having represented the (111) planes of the cube (4), as the base of the tetrahedron is situated at two-thirds of the height (or one-third from the base), which yields:

$$2a'\frac{\sqrt{3}}{3} \quad \text{or} \quad \frac{2}{\sqrt{3}}a' \quad \text{or} \quad \frac{2}{\sqrt{3}}r\sqrt{2} \quad \text{or} \quad \frac{2\sqrt{2}}{\sqrt{3}}r \qquad [2.10]$$

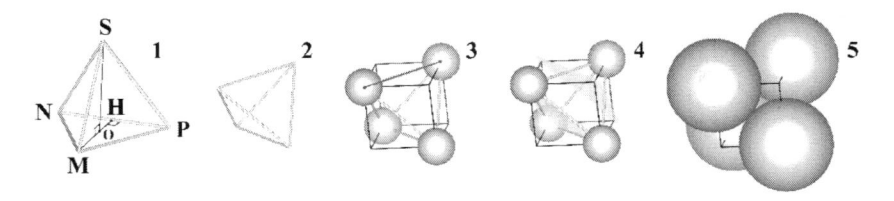

Figure 2.44. *Representations of a tetrahedron*

All these relations can be demonstrated in the tetrahedron (1) since the altitude OS, the base OM and the edge MS form a right triangle, and according to the Pythagoras theorem: $MS^2 = OM^2 + OS^2$. It should be kept in mind that in an equilateral triangle, altitudes, medians and perpendicular bisectors coincide and intersect at 2/3 of their length, which yields OM = 2/3 MH.

The length of MH is the altitude of the equilateral triangle:

$$MP \frac{\sqrt{3}}{2} \quad \text{or} \quad (2r)\frac{\sqrt{3}}{2} \quad \text{or} \quad r\sqrt{3} \qquad [2.11]$$

The length of OM is equal to 2/3 of MH, which is

$$\frac{2}{3}r\sqrt{3} \quad \text{or} \quad \frac{2}{\sqrt{3}}r \qquad [2.12]$$

The relation $MS^2 = OM^2 + OS^2$ or $OS^2 = MS^2 + OM^2$ can be used:

$$OS^2 = (2r)^2 - \left(\frac{2}{\sqrt{3}}r\right)^2 = 4r^2 - \frac{4}{3}r^2 = \frac{8}{3}r^2 \qquad [2.13]$$

And the root square yields $\frac{2\sqrt{2}}{\sqrt{3}}r$, as already shown previously [2.10].

2.3.2.3. *The hexagon*

The hexagon P has the unit cell parameters a and c and generally c > a and the atoms are in contact in the plane but not between planes (see Figure 2.45).

This case poses no difficulty; on the contrary, the hexagonal close-packed is worth focusing on (in this case, an atom appears in the half-plane of the conventional unit cell).

The ABA packing (1 and 2) confirmed by the anti-cuboctahedron (3) is visualized. It can be considered that packing A is represented by the four atoms and packing B by the central atom, in a bottom-up observation.

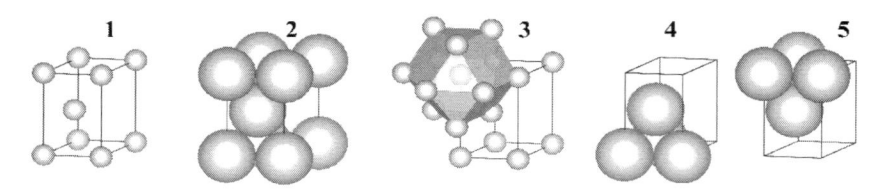

Figure 2.45. *Representations of a hexagonal close-packed structure*

Two tetrahedrons have been isolated (4 and 5), which are reversed and have a common atom, as a preliminary step to subsequent calculations.

The parameter c of this unit cell is therefore fixed at twice the altitude of tetrahedrons, which is $2\frac{\sqrt{2}}{\sqrt{3}}r$ (see [2.10]) or 1.633r and, since a = 2r, the parameter c is equal to 1.633 a.

The remaining calculations have already been presented in previous sections.

NOTE 2.12.– A type of lattice can be determined starting from measures of density and unit cell parameter, and vice versa. Indeed, the length of the edge of the unit cell depends on the radius of atoms (since they touch), the fraction of atoms in the unit cell can be calculated and this leads to the density of one unit cell.

For a small volume (as a is expressed in pm, a^3 expressed in cm^3 presents a factor of 10^{-30}), the mass is infinitesimal (it is the atomic weight divided by the Avogadro number in 10^{23}), the product of these factors yielding a coherent value in $g.cm^{-3}$.

2.3.3. *The sites*

The various packings described have free spaces that may contain smaller atoms, which are called interstitial sites. The atom placed in these sites has a specific environment, which is often tetrahedral (T) with four neighbors or octahedral (O) with six neighbors. The cubic F packing will be detailed, which has two types of sites.

2.3.3.1. *Octahedral sites*

When observing a cubic F packing (1), it can be noted (see Figure 2.46) that it is possible to place 13 small atoms in the spaces between the large atoms (2 and 3); these spaces are octahedral sites.

It is worth noting, however, that not all the interstitial atoms are present in the unit cell and that four interstitial atoms can be counted in the unit cell by the same method as the one applied previously for counting atoms (see section 2.2.3).

It can be noted that the small atom placed at the center of the cube (4) is surrounded by six large atoms (placed at the center of the faces of the cube) at identical distances ($a\sqrt{2}$), this atom being situated in an octahedral site.

All the atoms inserted are at the center of an octahedron but for good readability of the figure (5), only four of these octahedra have been represented.

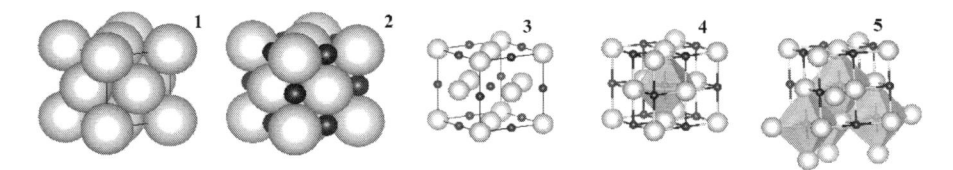

Figure 2.46. *Representations of octahedral sites in a cubic F packing. For a color version of this figure, see www.iste.co.uk/valls/inorganic.zip*

The maximum size of the small atom is given by the calculation of $\frac{a}{2} - r$ and as the diagonal of the face $a\sqrt{2} = 4r$ or $a = 2r\sqrt{2}$.

The result is $(\sqrt{2}-1)r$ and it can be proposed that the radius of the small atom should not exceed 0.414 times the large atom.

An octahedral site can be defined as a cavity situated at the center of a regular octahedron defined by six atoms in contact, which are in contact with the central atom and whose radii ratio is equal to 0.414.

2.3.3.2. Tetrahedral sites

Figure 2.47 shows a cubic F packing (1 and 2), in which it is possible to place eight small atoms in the spaces between large atoms (3 and 4), yielding a tetrahedral site.

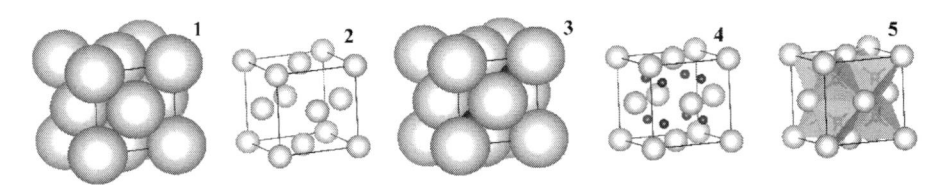

Figure 2.47. *Representations of tetrahedral sites in a cubic F packing. For a color version of this figure, see www.iste.co.uk/valls/inorganic.zip*

Each small atom is placed at the center of a tetrahedron and is surrounded by four large atoms at identical distances ($a\frac{\sqrt{3}}{4}$), this atom being situated in a tetrahedral site.

It can be noted that the eight atoms inserted are at the center of tetrahedra that have all been represented (5).

The maximum size of the small atom results from calculating $a\frac{\sqrt{3}}{4}-r$, and as it has been shown to be $a = 2r\sqrt{2}$ (see section 2.2.5.3), this yields $\left(\frac{\sqrt{3}}{\sqrt{2}}-1\right)$.

A tetrahedral site can be described as a cavity situated at the center of a regular tetrahedron defined by four atoms in contact, which are in contact with the central atom and whose radii ratio is equal to 0.225.

NOTE 2.13.– The conditions are identical for the hexagonal close-packed structure, as it should be remembered that both are cases of close-packed structures.

On the contrary, the sites in cubic I structures, represented in this figure (1 and 2) are pseudo-tetrahedral.

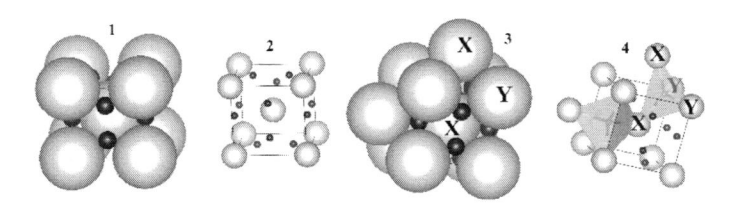

For a color version of this figure, see www.iste.co.uk/valls/inorganic.zip

For better readability purposes, only two sites have been represented (3 and 4) and if the XYXY tetrahedron is observed, it can be noted that the distance between atoms X and Y is a and the distance between X and Y is equal to the semi-diagonal of the cube or $\frac{a\sqrt{3}}{2}$. It is therefore a deformed tetrahedron.

2.4. Packings and diagrams

Now that the packings of identical atoms have been described, the behavior of various atoms in these packings can be explored. A choice has been made here to study only binary mixtures of metals in the form of alloys.

An alloy is different from a mixture of metals. For example, in a mixture of two powders, of gold and copper, for example, there is always gold and copper, no matter how fine the powder is. By contrast, in an alloy obtained by the crystallization of at least two metals brought to melting, the mixture of these two elements can be observed at atomic scale. The formed lattice, either regular or not, contains atoms of both metals (in this case gold and copper).

The solidification of a liquid mixture of two pure metals A and B is a very complex process, which can be analyzed by means of phase diagrams, an example of which is shown in Figure 2.48.

When two elements are molten together, the resulting material can be a single-phase alloy or a multi-phase alloy. Their number is linked to the intersolubility of elements, which is mainly dictated by their crystalline nature and the relative size of their atoms.

On the basis of a virtual binary phase diagram in A and B, various phases can be defined, as they appear during the cooling of a mixture of two compounds.

There are essentially three types of phases in an alloy, namely pure metals, intermetallic compounds and/or solid solutions:

– pure metals can emerge under certain conditions in the alloys and, for example, in zone 1 of Figure 2.48, pure compound A appears;

– intermetallic compounds can be found in the alloys, this term denoting a compound in which there is a simple proportion between the two elements and which is therefore similar to a chemical formula. For example, in zone 2 (a vertical line) of Figure 2.48, the compound denoted by AB_2 is present;

– solid solutions can be formed in the alloys and they are mixtures of atomic-scale elements, comparable to a mixture of mutually soluble liquids (which justifies the use of the word solution). In zone 4, there is a solid solution of A in B, knowing that on this diagram, B is insoluble in A. Intermediate compounds are present (zone 3) when the solid solution is preceded and/or followed by solid solutions named primary (zone 4).

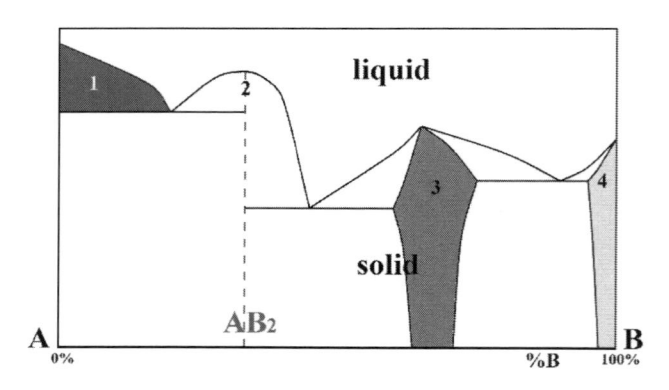

Figure 2.48. *Virtual diagram presenting all types of compounds. For a color version of this figure, see www.iste.co.uk/valls/inorganic.zip*

There are two types of solid solutions:

– substitutional solid solutions that contain in their lattice undesired replacement atoms (impurities or point defects) or voluntary replacement atoms (for low proportions of replacements, there are doped compounds or alloys for significant proportions);

– insertion solid solutions that can be formed of interstitial atoms (impurities or point defects) in low or significant quantities (there are additional elements).

It is worth noting that in both cases atom replacement is random and only the average composition of the solid solution can be given.

2.4.1. Reading the diagrams

In the previous example related to the binary gold–copper mixture, a continuous solid solution can be observed, to which a complete mutual solubility binary diagram can be associated (see Figure 2.49). This means that regardless of the gold/copper proportion, the diagram contains at least one solid and one corresponding liquid.

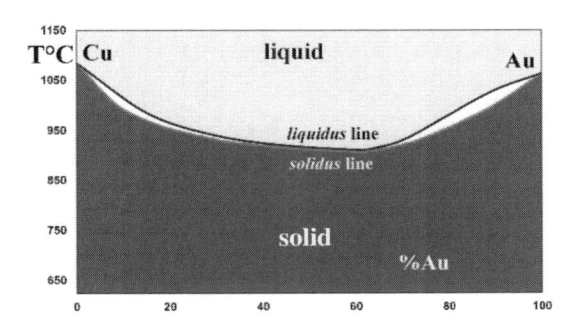

Figure 2.49. *Gold–copper binary diagram. For a color version of this figure, see www.iste.co.uk/valls/inorganic.zip*

This type of diagram is characterized by a double spindle limited by a *liquidus* line (upper line of the spindle) and a *solidus* line (lower line of the spindle). Gold–copper diagrams are difficult to use; therefore, a fictitious, more readable A-B diagram is instead proposed in Figure 2.50.

This type of diagram can be observed, for example, for alloys such as Au/Ag, Co/Ni, Sn/Bi, Cu/Ni, Bi/Sb and Cu/Ni.

2.4.1.1. Drawing a diagram

For a more detailed description, a simplified complete mutual solubility binary diagram (see Figure 2.50) has been drawn.

These diagrams should not be read like those of liquid–vapor equilibrium, although they are similar. In fact, due to viscosity, there is no separation between solid and liquid similar to the liquid–vapor separation. Two phases, a solid and a liquid one, intimately mixed but of different composition, coexist in the spindle.

In order to draw this type of diagram, known mixtures of A and B are successively brought to melting and their temperature evolution is observed during the cooling process (indicated on the right-hand side of Figure 2.50):

– The same type of plateau (1) is observed for each of the pure substances;

– When passing between the *liquidus* and *solidus* lines (across the spindle), all the mixtures present a constant slope (2 to 4), which is different from the previous one, as the solid crystallizes and releases heat, which slows the cooling and modifies the initial slope imposed by the furnace.

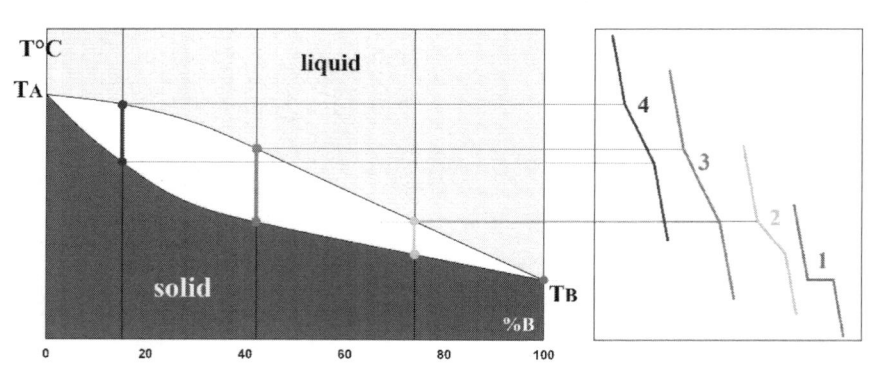

Figure 2.50. *Drawing a simple binary diagram. For a color version of this figure, see www.iste.co.uk/valls/inorganic.zip*

Drawing the lines corresponding to different compositions yields the full diagram thanks to their inflection points.

2.4.1.2. Reading a diagram along a horizontal line

This type of diagram can be read along a horizontal line and, if a solid having a B compound of composition X_N is brought to temperature T_N, then solid and liquid phases are intimately mixed throughout the spindle area.

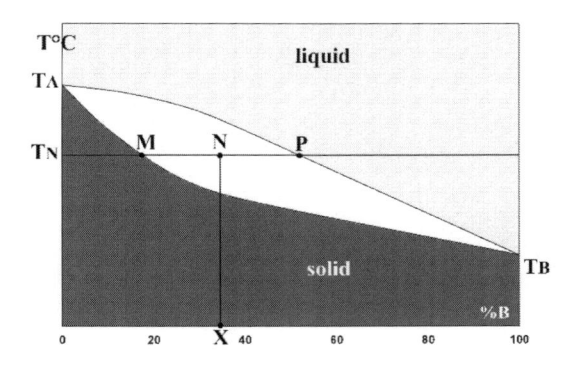

Figure 2.51. *Reading a simple binary diagram along a horizontal line*

When N tends to M, there is more and more liquid, as NP increases, and when N reaches M, only liquid is left.

In a simplified approach, the lever rule can be applied (see Figure 2.51) in order to determine the proportion of liquid and solid (expressed in percentage):

– The ratio of segment lengths $\frac{NP}{MP}$. 100 yields the percentage of solid;

– The ratio of segment lengths $\frac{MN}{MP}$. 100 yields the percentage of liquid.

NOTE 2.14.– The segment proportional to % of liquid (MN) is on the side of solid as, N being closer to P, there is more liquid. Conversely, the segment proportional to % of solid (NP) is on the side of liquid.

This is one of the consequences of the lever rule, as it is on the shorter side that there is a larger weight and vice versa.

The point of support being N, the closer it gets, for example, to the liquid (point P), the less solid there is, and NP can only be proportional to the quantity of solid.

2.4.1.3. Reading a diagram along a vertical line

Reading this type of diagram by moving along a vertical line (see Figure 2.52) means that two compounds are mixed in order to obtain a composition X that is liquid above point A and solid below point B.

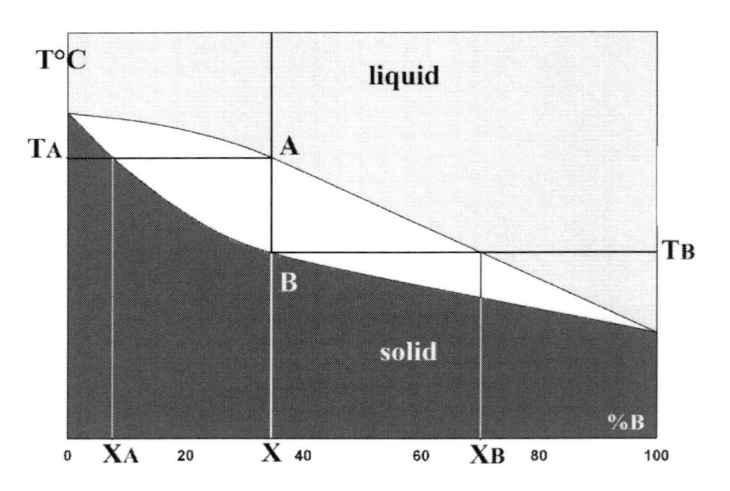

Figure 2.52. *Reading a simple binary diagram along a vertical line*

When temperature decreases:

– the first crystal grain of composition X_A is formed at point A;

– between A and B, there is a mixture of solid (of composition ranging between X and X_A) and liquid (whose composition ranges between X and X_B) and the lever rule can be applied to determine the proportions of solid and liquid at each temperature;

– at point B, the last liquid drop has the composition X_B;

– below point B, everything is solid and the overall composition is again X but, from a microscopic point of view, variable compositions can be observed depending on the cooling rate and these are due to successive crystallizations of compounds whose composition ranges from X_A to X.

Reading this type of diagram along a vertical line shows that temperature decreases and the composition of the emerging solid varies continuously.

To conclude this description, the macroscopic behavior of liquid transitioning to crystal can be represented as below (see Figure 2.53).

Figure 2.53. *Macroscopic structure of metal during liquid-to-crystal transition*

A pure metal or a complete solubility alloy can solidify from a liquid phase (1) starting with nucleation points (2) where solid grains are formed and will grow (2 and 3) until they come into contact (4) and form a solid phase. These are polycrystalline solids whose grains are separated by grain joints represented by lines (4).

A crystalline solid can be obtained by annealing, which is heating to an adapted temperature and for an adapted period of time so that atoms can rearrange and get ordered (5).

The crystals that are described here are considered without grains or grain joints as well as without defects (perfect crystal model).

2.4.2. Solid solutions

In solid solutions, the structure of one of the components is preserved (with several modifications) and atoms of the other component occupy sites that are normally occupied by atoms of the first metal.

2.4.2.1. Substitutional solid solutions

Gold and copper crystallize in cubic F system and are miscible in all proportions. Crystal atoms are placed in sites that have homogeneous distribution (see Figure 2.54), either irregular (1) of one metal into the other or regular (2), which will be detailed in section 2.4.3. If the atoms are represented along a plane, the random distribution (1) will be visualized, which always has the same proportion irrespective of the plane (here it is close to 50/50).

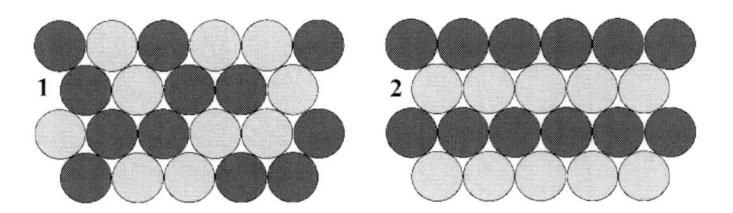

Figure 2.54. *Plane representation of a solid solution of gold and copper*

Five empirical conditions can be defined for the existence of substitutional solid solutions:

– crystalline structures of the same type feature high solubility, in contrast with structures of different type;

– solubility is inversely proportional to the size difference between atoms (there is limited solubility beyond 15% size difference);

– being part of the same classification block is one of the solubility conditions (copper and nickel or gold have complete solubility, whereas copper and tin have only partial solubility below 20%);

– metals cannot randomly substitute for one another unless the difference in electronegativity between them is sufficiently small;

– metals of the same valence are readily dissolved and the higher the difference in valence, the lower their solubility.

If all the factors are favorable, complete solubility is most often observed, which means that there is full miscibility for a solid solution irrespective of proportions. In the opposite case, there are limited solubility domains.

The rules are necessary but not sufficient; the fact that all rules are satisfied does not guarantee that the compounds are soluble and, for example, copper and silver have only partial solubility, whereas copper and gold as well as silver and gold feature full solubility.

2.4.2.2. *Insertion solid solutions*

Insertion solid solutions are obtained when crystal gaps are occupied by small atoms, generally hydrogen, boron, carbon and nitrogen, knowing that only hydrogen can occupy small tetrahedral sites. These atoms inhabit a site of the host lattice without generating excessively significant deformations.

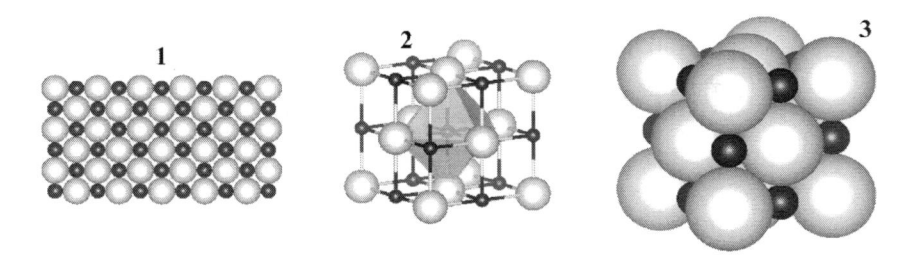

Figure 2.55. *Representations of intermetallic compounds. For a color version of this figure, see www.iste.co.uk/valls/inorganic.zip*

The most frequent insertion alloys emerge on face-centered cubic packings, the insertion atoms being arranged in the octahedral sites of this structure.

For example, tungsten carbide is a very resistant alloy used in cutting tools; it results from the insertion of carbon atoms in all the octahedral sites of the tungsten lattice. The presence of carbon prevents tungsten atoms from relative slipping and the compound becomes very hard as well as more brittle.

The most common example is that of steels, which are insertion solid solutions of carbon in iron. The solvent is represented by iron atoms and the solute by carbon atoms inserted between iron atoms. A representation of the limit solid solution has been drawn (see Figure 2.55), in which all sites are occupied (1 and 3) and the octahedral site is occupied by the carbon atom at the center of the unit cell (2).

Steels correspond to partial and generally random but homogeneous occupations of these sites.

> NOTE 2.15.– The diagram of steels is always surprising, as it is most often marked in carbon percentage, which does not exceed 5%. Indeed, it is a weight fraction, which is smaller than the mole fraction.
>
> The molar mass of carbon is 12 and that of iron is 55.8; therefore, a 100% mole-to-mole mixture will only have a weight fraction of 17.7%. This explains the low proportion of carbon when expressed in a weight fraction.

2.4.3. Intermetallic compounds

Although the notion of valence has no significance for intermetallic compounds, they can be considered genuine chemical compounds.

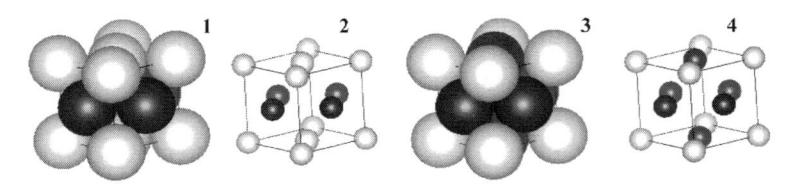

Figure 2.56. *Representations of ordered solid solutions $AuCu_3$ and $AuCu$*

Gold and copper are miscible in all proportions and, for mole–to-mole (AuCu) and one mole to three moles ($AuCu_3$) compositions, ordered phases can be obtained (see Figure 2.56). The atoms are distributed in accordance with symmetries and this leads to solids that are perfectly characterized, for which definite formulae can be proposed: AuCu (1 and 2) and $AuCu_3$ (3 and 4).

Ordered solutions have symmetries that are different from those of pure solids because of alternating atoms (see Figure 2.56). For example, gold and copper have cubic F structure, and ordered solid solutions have tetragonal symmetry (1 and 2) and cubic symmetry (3 and 4).

It can be implied that they have different properties, since the first one has a layered structure and the second one is isotropic, as the cube formed by gold has the same symmetries as the octahedron formed by copper atoms.

The proportion of gold and copper in AuCu is 50% gold atoms and 50% copper atoms and, if the composition does not change, then, above 450°C, gold and copper atoms are randomly distributed and packing symmetries are lost. This transition from ordered to disordered state takes place when temperature exceeds a value that is characteristic of the type of metal.

2.4.4. Simple phase diagrams

2.4.4.1. Diagrams for total miscibility

The representation of diagrams for total miscibility highlights two types: the ones with simple spindle and those with double spindle.

A fictitious diagram is shown in Figure 2.57. It presents a minimum (with M below T_A and T_B) but it can also present a maximum (with M above T_A and T_B).

Two phases (solid and liquid) can be observed in the spindles, while above and below them there is only one phase, liquid and solid, respectively.

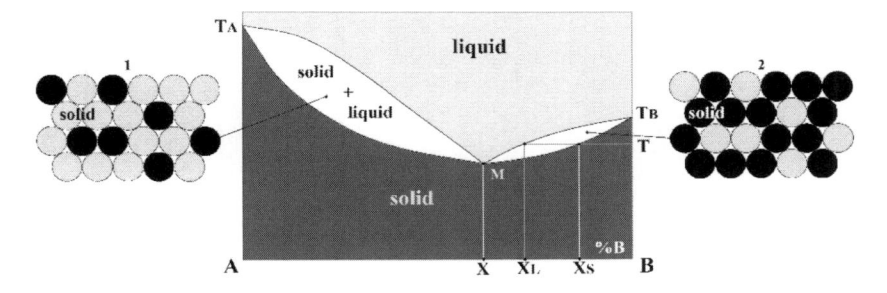

Figure 2.57. *Diagrams for total miscibility with a minimum*

These diagrams are obtained by the method described in the previous section (see section 2.4.1.1) by bringing known mixtures of A and B to melting and observing the evolution of temperature during their cooling.

Four areas of this fictitious diagram for total miscibility with minimum can be evidenced (see Figure 2.57).

At high temperature (light blue area), homogeneous liquid is obtained regardless of the composition because miscibility is linked to the liquid state. Similarly, at low temperature (dark blue area), a homogeneous solid is obtained, regardless of the composition.

The two spindles feature mixtures of solid and liquid. In the spindle on the left-hand side, the solid is a solid solution (1) of B (black atoms) in A (gray atoms) and in the spindle on the right-hand side, it is a solid solution (2) of A in B.

114   Inorganic Chemistry

The proportions of solid and liquid are given by the lever rule (see section 2.4.1.3) and the solid and liquid compositions at temperature T are obtained by the intersection of a horizontal line drawn at temperature T, with the *solidus* line (for solid X_S) and with the *liquidus* line (for liquid X_L), respectively.

The mixture of A and B of composition X, which corresponds to the minimum (M), melts or solidifies as a pure substance, at constant temperature.

Diffusion in solid phase is also weak, the resulting solid taking the form of layer around a grain (first solid formed) with a composition gradient, but the average composition of the solid is necessarily X.

The lower part of the diagram corresponds to a solid that is generally unordered but annealing makes it possible for certain alloys to form ordered phases.

2.4.4.2. Eutectic diagrams with non-miscibility

Diagrams with non-miscibility may feature a specific point (E), which marks the composition of a mixture qualified as eutectic (see Figure 2.58).

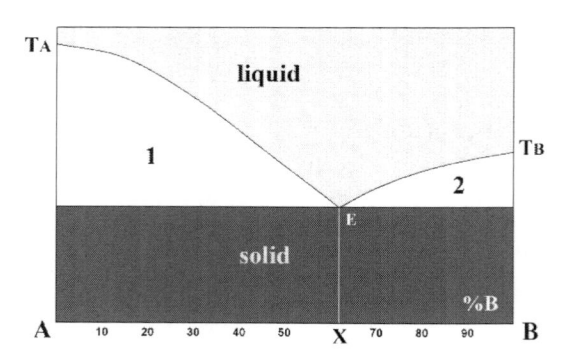

Figure 2.58. *Eutectic diagrams with non-miscibility*

The upper part of the diagram, above the *liquidus* line, defines the area of existence of a liquid (light blue).

The area under the horizontal line that passes through point E is defined as the area of existence of a homogeneous solid (in dark blue).

At the eutectic point (E), the solid phase has the same composition as the liquid phase and the eutectic mixture behaves similarly to the pure substance and melts or solidifies at fixed temperature.

Area (1) corresponds to a biphasic mixture in which the solid is compound A and the liquid is a homogeneous mixture of A and B, which becomes scarce in A when temperature decreases.

Symmetrically, area (2) corresponds to a biphasic mixture in which the solid is compound B and the liquid is a homogeneous mixture of A and B, which becomes scarce in B when temperature decreases.

For these two areas, the last fraction of liquid before solidification is the eutectic of composition X.

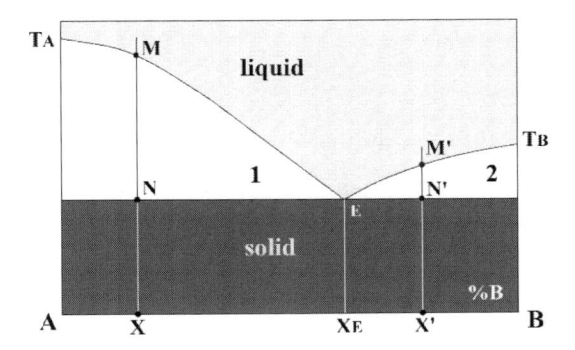

Figure 2.59. *Various areas of a non-miscibility diagram with eutectic point*

The diagram in Figure 2.59 can be detailed, by moving along vertical lines (MN then M'N'):

– a homogeneous liquid can be observed above M; then, at point M, the first solid grain emerges, which is pure A compound. Crystals of pure A compound continue to grow from M to N and the liquid is enriched in B up to N;

– the whole becomes solid at point N and its composition is X from a macroscopic point of view.

Observation of the solid leads to identifying pure A solid surrounded by eutectic of composition X_E, the whole being of composition X.

Symmetrically, from M' to N', the solid that emerges is pure B compound and, at point N', the whole becomes solid and the composition is X' from a macroscopic point of view. The observation of the solid leads to identifying pure B solid surrounded by eutectic of composition X_E.

In both cases, the solid is therefore heterogeneous from a microscopic point of view.

> NOTE 2.16.– The best-known non-miscibility diagram with eutectic is that of ice and salt; in winter, roads are salted in order to take advantage of the properties of eutectic, which is liquid up to −21.6°C (it should be noted that the liquid is salted water!).
>
> Up to −21.6°C, ice floats in salted water (provided there is enough salt), forming a slush that covers the roads when it is not snowing heavily. However, when the temperature drops below −21.6°C, the tracks carved by vehicles in the slush become rails, everything freezes and road traffic becomes difficult. For lower temperatures, calcium chloride is used, as it features a eutectic liquid point of up to −51.1°C.

2.4.4.3. Diagrams with partial miscibility

For certain binary systems in solid phase, the constituents are not miscible in all proportions and there is a partial solubility that is generally linked to the difference in the type of lattice between the two compounds.

In this case, the diagram of partial miscibility has a similar form to the diagram shown in Figure 2.60 with monophasic areas (1, 3 and 4) and biphasic areas (2, 5 and 6). The upper part of the diagram (1), which is above the *liquidus* line, defines the area of liquid.

Below the horizontal line passing through point E (2), the area of solid composed of two phases is defined. It is a solid solution of B in A (3) and another one of A in B (4), whose proportion is defined by the lever rule and composition by the horizontal line at the desired temperature.

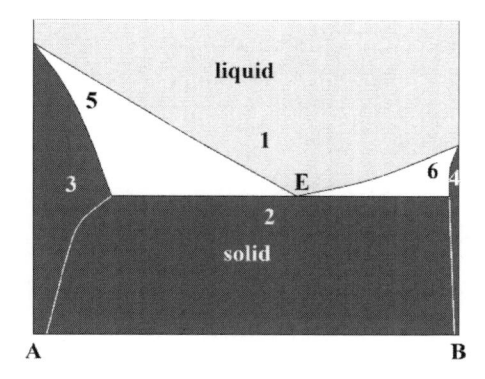

Figure 2.60. *Various areas of a diagram with partial miscibility*

Areas 5 and 6 correspond to mixtures of solid and liquid whose proportions and composition are determined by the usual methods (lever rule and horizontal line).

At the eutectic point (E), the solid phase has the same composition as the liquid phase and the eutectic mixture shows the same behavior as a pure substance, that is, it melts or solidifies at fixed temperature. The formed solid is biphasic since it is in area (2) and contains two solid solutions of B in A and A in B.

A detailed study of this diagram is presented in section 2.4.4.5.

2.4.4.4. *Presentation of various types of diagrams*

The observation of various diagrams shows that it is often possible to decompose them into simple diagrams.

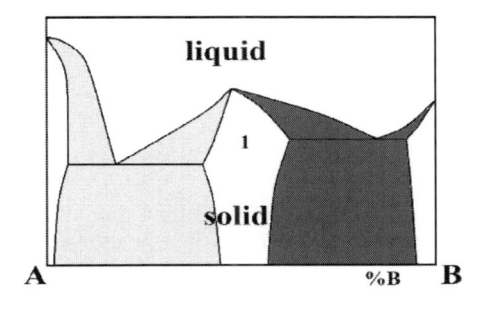

Figure 2.61. *Representation of a diagram with intermediate compound*

The first example of a fictitious diagram proposed (see Figure 2.61) presents in light blue a diagram with partial miscibility and in dark blue a second one beside it, the two being separated by an intermediate solid solution (1) with congruent melting.

The second example proposed (see Figure 2.62) shows in dark blue the left-hand side of a diagram with partial miscibility (1) and the right-hand side of another diagram with partial miscibility (2).

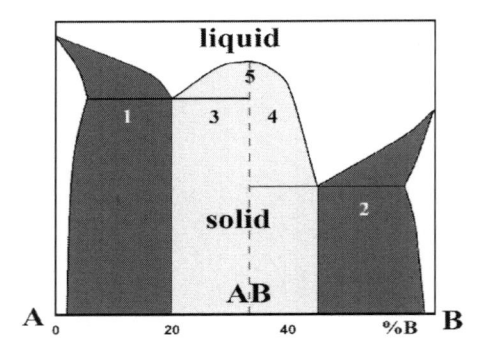

Figure 2.62. *Representation of a diagram with partial miscibility with congruent melting compound*

The central part contains the right-hand side of a non-miscibility diagram (3) and the left-hand side of a non-miscibility diagram (4).

An intermetallic congruent melting compound can be visualized at the center, its formula being AB (5).

2.4.4.5. Detailed study of a diagram

The diagram of lead and tin is simple and can be used in order to readily analyze the various phenomena that take place in a binary mixture during cooling. It will therefore be used for a detailed description of the macroscopic structure of the solid obtained.

Lead has a cubic F structure with a lattice parameter a equal to 495.1 pm and tin has a tetragonal structure with a lattice parameter a equal to 583.2 pm and c equal to 318.1 pm. Zero or weak miscibility are therefore expected.

Indeed, the diagram indicates that tin is soluble up to 19% in lead (point N) and that lead is soluble up to 2.5% in tin (point P), as expressed in molar percentages.

The observation of the lead–tin alloy equilibrium diagram indicates the existence of five areas, as shown in Figure 2.63, from 0 to 19% tin (1), which is the limit of solubility of tin in lead; from 19 to 61.9% (2); the 61.9% composition presenting specific properties (3); from 61.9 to 97.5% (4); and from 97.5 to 100% (5), which is fixed by the limit of solubility of lead in tin (2.5%).

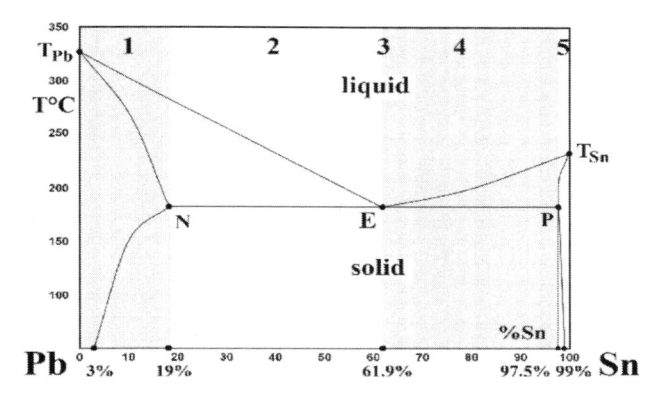

Figure 2.63. *Diagram of lead–tin binary alloy*

A description of various changes taking place between melting and complete solidification in composition and macroscopic structure, depending on temperature, is shown in the figure.

In Figure 2.63, lead-melting temperature is (T_{Pb}) and tin-melting temperature is (T_{Sn}), whereas the eutectic point is (E). The figure also shows various values of composition (in tin molar percentage) for all specific points.

Following a vertical line in the first area (see Figure 2.64), for a mixture of composition X (arbitrary between 3 and 19% of tin), the aspects below can be noted:

– above A, there is liquid (of composition X) and then the first solid grains appear in A (composition X_A), which then grow and their composition changes from X_A to X, up to B, while the composition of the liquid goes from X to X_B;

– between B and C, there is an unordered solid solution of tin in lead (dark blue, of composition X);

– below C, the solid is composed of two phases, which are unordered solid solutions, one rich in lead (dark blue, evolving from X to 3%) and the other one rich in tin (medium blue, evolving from X_C to 99%). When this second phase rich in tin is formed, the first becomes poor in tin and a heterogeneous solid can be observed, in which the two solids are intertwined in a more or less homogeneous manner.

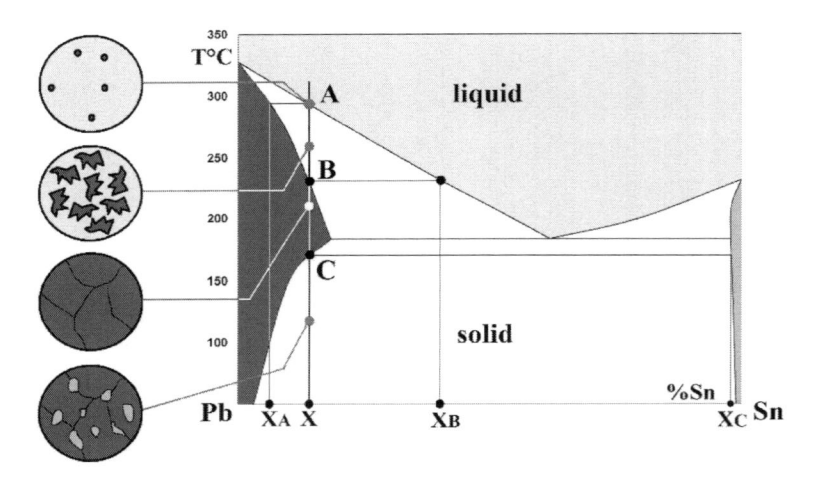

Figure 2.64. *Diagram of lead–tin binary alloy (area 1). For a color version of this figure, see www.iste.co.uk/valls/inorganic.zip*

Proportions and composition can be determined by the usual methods (drawing a horizontal line and lever rule).

Following a vertical line in the second area (see Figure 2.65) for a mixture of composition X (arbitrary between 19 and 61.9% of tin), the aspects below can be noted:

– above D, there is liquid (of composition X) and then the first solid grains are formed in D (of composition X_D), which then grow up to F (composition varying from X_D to 19%). The last fraction of liquid has a composition of 61.9% in tin;

– above F, the liquid becomes poor in lead until it reaches the eutectic composition;

– below F, the whole gets solidified and the solid is composed of crystals of unordered solid solutions that are rich in lead (in dark blue, containing 19% tin) surrounded by solid eutectic generally in laminae. Eutectic laminae are composed of two unordered solid solutions of tin in lead (dark blue, whose composition evolves from 19 to 3% in tin) and of lead in tin (medium blue, which evolves from 97.5 to 99% in tin).

Proportions and composition are determined by the usual methods.

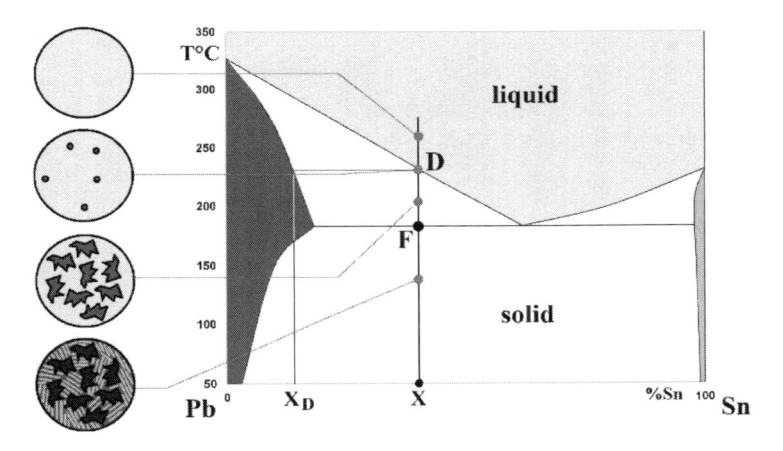

Figure 2.65. *Diagram of lead–tin binary alloy (area 2). For a color version of this figure, see www.iste.co.uk/valls/inorganic.zip*

The third area in Figure 2.66 is identified by point E:

– above E, there is homogeneous liquid of composition X_E;

– below E, the solid has the same composition X_E but its structure is heterogeneous, with a mixture of two unordered solid solutions of tin in lead and lead in tin, which are generally structured in alternate laminae, as described previously.

The behavior of this mixture is similar to that of a pure substance. Compositions and proportions are determined by the usual methods.

The fourth area starts above G with a liquid (see Figure 2.66):

– starting from G, the first solid grains emerge (of composition X_G) and then grow (composition ranging between X_G and 97.5%) up to H;

– below G, the liquid becomes poor in lead and the last fraction of liquid has eutectic composition;

– below H, the whole gets solidified and the solid is composed of crystals of unordered solid solutions rich in tin (in medium blue, containing 97.5% tin) surrounded by solidified eutectic, generally in laminae.

The behavior of areas 2 and 4 is similar, knowing that solid solutions are swapped. In this case, a solid solution of lead in tin emerges and it freezes due to the solidification of eutectic laminae when temperature decreases. Proportions and composition can be determined by the usual methods (drawing a horizontal line and lever rule).

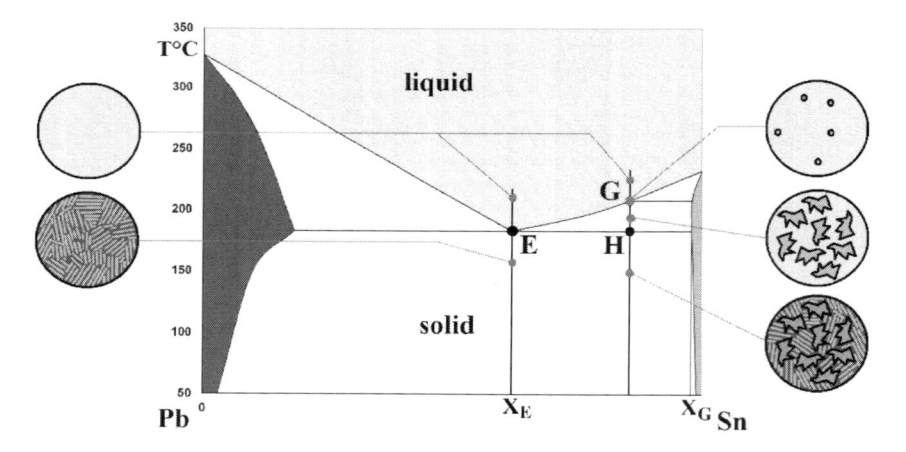

Figure 2.66. *Diagram of lead–tin binary alloy (areas 3 and 4). For a color version of this figure, see www.iste.co.uk/valls/inorganic.zip*

Following a vertical line in the fifth area (see Figure 2.67) for a mixture of composition X (arbitrary between 97.5 and 100% of tin), the aspects below can be noted:

– above K, there is a liquid of composition X, and then in K, the first solid grains emerge with a composition X_K and they grow up to L (composition ranging from X_K to X), while the composition of liquid evolves from X to X_L;

– between L and M, there is an unordered solid solution of lead in tin (medium blue, of composition X);

– below M, the solid is composed of two phases that are unordered solid solutions, one rich in tin (medium blue, which evolves from X to 99%) and the other one rich in lead (dark blue, which evolves from X_M to 3%).

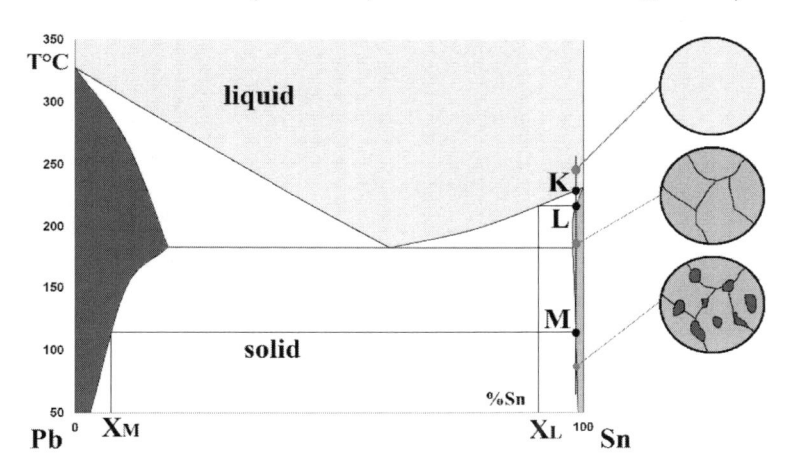

Figure 2.67. *Diagram of lead–tin binary alloy (area 5). For a color version of this figure, see www.iste.co.uk/valls/inorganic.zip*

When this second phase rich in lead is formed, the first one gets scarce in lead and a heterogeneous solid is observed, with two intertwined phases similar to area 1, but solid solutions are permutated.

Proportions and composition can be determined by the usual methods (drawing a horizontal line and lever rule).

3

Knowledge of Ionic Crystals

3.1. Description of ionic to covalent crystals

The structures of ionic crystals are much more diversified than those of metallic crystals, as they do not involve atoms of the same size and are structured by an isotropic bond of metallic type. Indeed, ions have sizes that vary more widely than those of metals, they can carry one or more positive or negative charges and their bonds cover a very wide range.

Crystalline solids can therefore be metallic solids, ionic solids such as salt, as well as covalent solids such as diamond, molecular solids such as dry ice or solids structured by hydrogen bonds, such as ice, or by Van der Waals forces, such as graphite or, still, intermetallic compounds that involve a more or less metallic type of bond. Consequently, the crystal can be defined by metallic, ionic, covalent and/or Van der Waals radii.

This diversity of bonds confers multiple properties to crystalline compounds and accounts for the richness of this field of chemistry.

Any classification of bonds remains an academic endeavor, as there is no one theory on the nature of bonds. Indeed, there are substances that can make the transition from covalent to ionic or metallic character depending on temperature, pressure or type of structure. A simple to complex approach will be adopted here (to classifications, descriptions or representations) and several keys stemming from fundamental theories of solids will be used.

The perfect crystal model is therefore used for ionic crystals, as it was previously used for metals, and the ionic crystal is considered an ordered, infinite and regular packing of ions, devoid of any defects.

The hard sphere model is too simple to be applied to ionic crystals and more elaborate models consider ions to be composed of two parts: a hard central core that contains most of the electronic density and an outer soft sphere that is more or less polarizable and has a lower electronic density.

While this model can be readily imagined, the laws establishing the size of this soft part are currently more difficult to formulate.

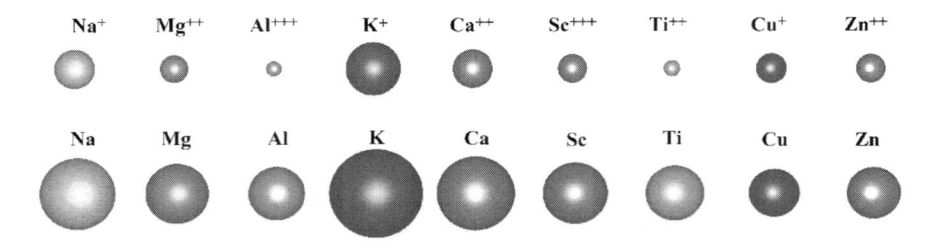

Figure 3.1. *Relative size of several (six-coordinate) atoms and ions. For a color version of this figure, see www.iste.co.uk/valls/inorganic.zip*

The ions (see Figure 3.1) are not hard spheres and their size is not constant, as it depends on their environment.

The ionic bond refers essentially to a combination of elements that have:

– few electrons in their outer shell (columns 1 and 2 of the periodic table) and will readily lose the one or several outer-shell electrons forming cations isoelectronic with the rare gas preceding them;

– quasi-saturated outer shells (columns 16 and 17 of the periodic table) that will gain one or several outer-shell electrons in order to saturate the valence shells and form anions isoelectronic with the following rare gas.

This electrostatic bond is strong, as its enthalpy of formation is of several hundred kilojoules per mole.

The materials structured by ionic bonds are characterized by:

– low thermal conductivity due to lack of electron mobility;

– good transparency, as no electron transition can absorb visible-light wavelengths;

– a certain brittleness and a high melting point.

When crystals have properties that differ from these characteristics, the type of bond that structures them can be considered to have evolved toward covalence (loss of transparency, for example) or metallic character (increase in conductivity).

It can be posited that the higher the formal charge of a metallic ion is, the more significant the degree of covalence of the bond with its neighbors and the higher the difficulty of applying the ionic radius concept.

NOTE 3.1.– In order to simplify the descriptions of crystals, reference or typical compounds have been defined.

For example, when a crystal is said to be of NaCl type, it can be concluded that anions and cations are distributed according to cubic F packings, where each ion is at the center of an octahedron formed of ions of opposite sign and an alternation of ions can be observed in all three space directions, etc. All these characteristics of NaCl crystal are identical with those of all its compounds said to be isomorphous (same structure, with the possibility of mixed crystals) or isotypic (same structure without the possibility of mixed crystals).

Here, the focus will be on five typical binary crystals whose formula is MX (CsCl, NaCl, ZnS sphalerite and wurtzite and NiAs) and other five crystals of formula MX_2 (CaF_2, Li_2O, $CdCl_2$, CdI_2 and TiO_2), as well as on two ternary crystals belonging to the family of perovskites of formula ABX_3 ($SrTiO_3$) and on spinels of formula AB_2X_4 ($MgAl_2O_4$). Several examples of specific compounds will be given in order to illustrate specific properties (ReO_3 [DYU 88], Cu_2O [KIR 90], $KCu_{0.3}Mg_{0.7}F_3$ [BUR 96], $CrMnNiO_4$ [REN 72], etc.).

It will be possible to apply the knowledge acquired on metallic packings, as the larger ions (generally anions) are organized according to models applicable to atoms in metals. The smaller opposite ions glide into interstitial sites created between the largest ions, all this leading generally to maximum packing density and overall neutral charge in the smallest possible volume.

The notions of packings, sites and coordination polyhedra will be reviewed, and the notions of charge and electrical neutrality will be explored. The hard sphere model, which is useful for metals, will be employed in an adapted version, as in ionic solids the size of ions depends on their environment. Nevertheless, once the size is defined, the hard sphere model can be used in the reasoning.

Since the focus is on crystals, a brief history of data related to this subject can give an idea of how recent they are.

This history can be considered to have started in the 17th Century, when the invention of the microscope made possible the observation of crystals. After only two centuries, it was discovered that different crystals may have the same form, a characteristic known as isomorphism, for example between salt or halite (NaCl) and galena (PbS). A further discovery was that crystals of different forms can have the same composition, which is known as polymorphism, for example between calcite and aragonite ($CaCO_3$) although data related to internal crystal structure was not yet available.

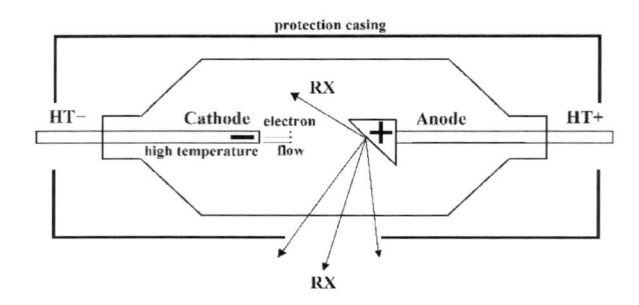

Figure 3.2. *Schematic representation of an X-ray tube*

The discovery of X-ray diffraction at the beginning of the 20th Century (preceded by the discovery of the X-ray tube, see Figure 3.2) has facilitated the

Knowledge of Ionic Crystals   129

study of the crystal's internal structure. However, it was only in 1960 when computers could be used for the first time for the characterization of crystalline structures that research on crystals took off.

By the end of the 20th Century, data on over 200,000 crystalline structures had been recorded in the databases and there are currently dozens of specialized databases that contain information on 4,000 to 400,000 substances.

This book draws on various databases, particularly on the freely accessible *Crystallography Open Database* (http://www.crystallography.net/), which gathers data on over 376,000 substances. The use of open sources plays an important role in the effective sharing of knowledge and is highlighted in various sections of this chapter.

3.2. Pauling's rules

Pauling proposed various rules that should be used with caution but can nevertheless serve as verification tools [PAU 29].

The simplified ionic theory can be used to explain the formation of coordination polyhedra. Given the hypothesis that atoms are non-deformable spheres and ions of opposite signs are in contact, the coordination number can be determined by simple geometrical considerations.

The structural principles posited by Goldschmidt for ionic crystals were processed by Pauling into a series of five rules that state electrical and geometric criteria for ionic solids. They have been successfully expanded to ionic covalent crystals, with some precautions.

These five rules are applicable to crystals with ideal structure, in which the ionic bond predominates:

– The first rule proposes geometric criteria and refers to the coordination polyhedra of cations. It specifies that the cation–anion distance is equal to the sum of ionic radii and the ratio of ionic radii determines the type of polyhedron. For example, Figure 3.3. shows representations of a cubic polyhedron (1 and 2), a tetrahedron (3 and 4) and an octahedron (5 and 6).

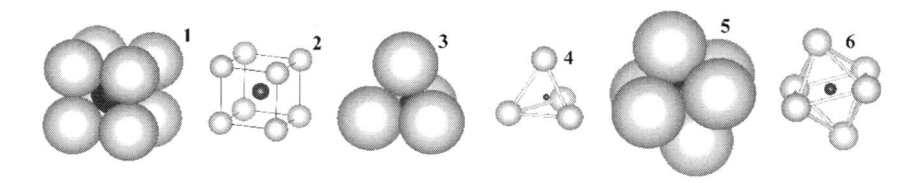

Figure 3.3. *Polyhedra of cubic, tetrahedral and octahedral coordination*

– The second rule refers to the electrostatic valence principle, knowing that nature imposes overall neutrality of the crystal as well as neutrality in the smallest possible volume.

– The other rules specify the connection of the polyhedra of anionic coordination, state the possibility of polyhedra forming independently (phosphate ion, sulfate ion, etc.) and postulate the homogeneity of the environment of ions. Nevertheless, there are numerous exceptions to these rules.

The first two rules, which are generally followed, will be commented upon as a priority for each type of crystal described; the rest of the rules will be mentioned only if they facilitate the understanding of structures.

3.2.1. The ionic character of a bond according to Pauling

Pauling calculated the ionic character of a bond, with quantity expressed as a percentage, that will subsequently be referred to as theoretical ionicity (or simply ionicity), between two atoms A-B starting from electronegativities of A (χ_A) and B (χ_B), using the following formula:

$$\text{ionicity} = 100 \left[1 - e^{\frac{(\chi_A - \chi_B)^2}{4}} \right] \qquad [3.1]$$

If theoretical ionicity tends to 0%, the bond is predominantly covalent; if it tends to 100%, the bond is predominantly ionic and if it ranges between these two values, the bond is ionic covalent.

Fully covalent bonds are characteristic of simple substances (H_2, O_2, F_2, etc.), whereas fully ionic bonds do not exist and the maximum theoretical ionicity can be calculated on the basis of extreme electronegativities: fluorine ($\chi = 3.98$) and francium ($\chi = 0.7$), which yields 93%.

If this is applied to the periodic table (see Figure 3.4), four areas are evidenced for metals:

– most of the metals have $\chi \approx 1.8$;

– very reducing metals have $\chi \leq 1$ (alkaline and alkaline-earth metals, except Mg and Be);

– reducing metals have $\chi \approx 1.4$;

– specific metals have $\chi > 2$ (mainly the catalysts).

In Figure 3.4, the boxes corresponding to metals whose electronegativity is above 2 are white and those corresponding to synthetic elements are struck-through. The non-metals yielding insertion compounds (H, C, N and P) are also highlighted, as well as the set of metalloids.

Non-metals form ionic bonds with most of the metals (NaF, $MgCl_2$, AgI_2, NaBr, etc.) but hydrogen can only form ionic bonds with alkaline metals and yields ionic covalent hydrides (NaH, LiH, etc.).

Non-metals form only covalent bonds with one another (Cl_2, N_2, PCl_3, IF_5, NH_3, HCl) irrespective of the difference in electronegativity between them.

Metals are exclusively donors, so that if a metal engages in a (non-metallic) bond, it is always a donor and never an acceptor. The degree of oxidation of metals is therefore always positive or null, but never negative.

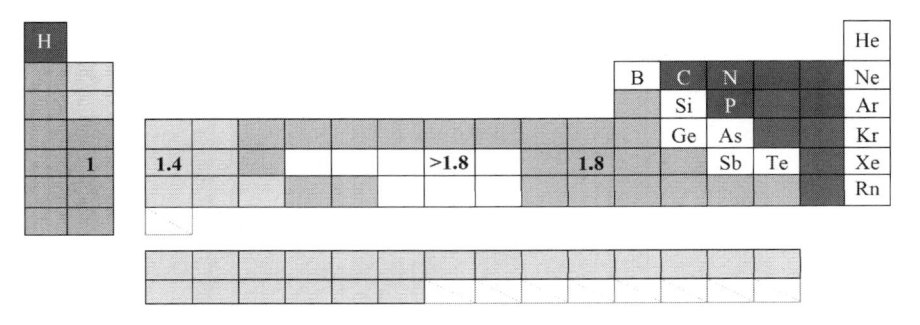

Figure 3.4. *Evolution of electronegativity in the periodic classification. For a color version of this figure, see www.iste.co.uk/valls/inorganic.zip*

Metals do not chemically react with each other regardless of the difference in electronegativity and their atoms get packed and form alloys. They can yield organized crystalline compounds under certain conditions of temperature and composition (see section 2.4.3).

Non-metals and metals generally yield ionic covalent compounds, knowing that H, C, N and P can yield insertion compounds; that O, F and Cl yield mainly ionic compounds and that S, Se, Br, I and At yield mainly covalent bonds.

Metalloids combine with metals and non-metals and generally yield semiconductors.

If two atoms A and B form a crystalline compound AB, the type of bond in this crystal can be predicted on the basis of the difference in electronegativity ($\Delta\chi$) and therefore from the position of the elements in the periodical table (see Table 3.1).

In general, the classifications propose Pauling's values of electronegativity, which have no (or little) variation when the charge of ions varies.

For several elements, electronegativity increases significantly with the state of oxidation (Ag, Mn, Pb, P and Sn) but, for convenience, only Pauling values will be considered here, as they are sufficient for the objective of this study of crystals.

Nature of elements A and B	$\Delta\chi$ or compounds	Type of bond AB
Non-metals	molecules	always covalent
Non-metals and metals	if $\Delta\chi \leq 0.5$	strongly covalent
	if $0.5 \leq \Delta\chi \leq 1.9$	ionic covalent
	if $\Delta\chi \geq 1.9$	strongly ionic
Metals	alloys	always metallic

Table 3.1. *Type of bond depending on the nature and electronegativity of elements*

3.2.2. Pauling's first rule: coordinated polyhedra

When spheres of different sizes are packed, the ratio of the central cation radius to surrounding anion radius makes it possible to predict the simple volumes, if the spheres are supposed to be in contact.

The cation–anion distance is determined by the sum of ionic radii, and in this first Pauling's rule, the coordination number is defined as the ratio of cation radius (R^+) to anion radius (R^-), denoted by R^+/R^-.

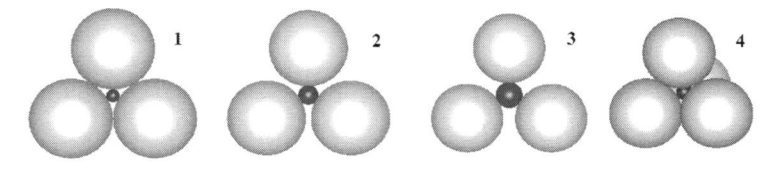

Figure 3.5. *Coordinated polyhedra with three or four anions*

Considering the example of ideal coordination number 3 (1), in which all ions are in contact (see Figure 3.5), when the size of cation increases (2 and 3), the coordination number is generally preserved until the ratio R^+/R^- equals that of a higher coordination number (2). Beyond (3), the coordination number changes and, in this case, tetrahedral coordination is achieved (4).

It should nevertheless be considered that stability is the highest when the coordination of a cation is the highest possible, while observing the cation–cation contact criterion. Several geometry elements make it possible to prove the existence of a link between coordination and the ratio of ions' radii (R^+/R^-), which will be referred to as steric criterion.

NOTE 3.2.– Valence has no direct impact on the coordination number. If NaCl and PbS structures are considered, they are of similar type (NaCl) despite the evolution of cation valence, which varies from 1 to 2, and the ionic character of bonds, which varies from 71 to 15%. The same is applicable to BaF_2 and PbO_2, which are of CaF_2 type, whose valences vary from 2 to 4 and ionicity from 91 to 51%.

Hasty generalizations are not recommended, as NpO_2 and VO_2 have similar formulae, ionicities and valences but different structures (CaF_2 and TiO_2 type) with coordination numbers 8 and 6, respectively.

3.2.2.1. Coordination number 4

In general, in four-coordinate systems, four anions form a tetrahedron (1 and 2) at the center of which a cation glides (see Figure 3.6).

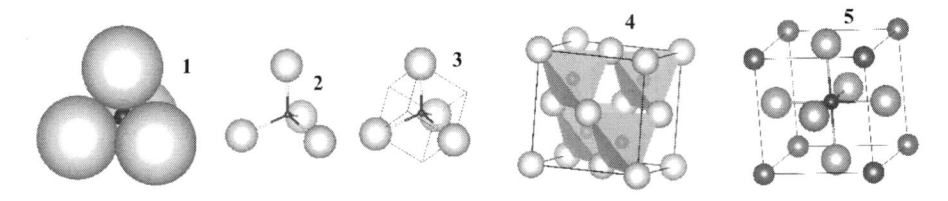

Figure 3.6. *Examples of four- and six-coordinate structures (ZnS sphalerite and SrTiO₃). For a color version of this figure, see www.iste.co.uk/valls/inorganic.zip*

The tetrahedron is placed in a cube (3) whose face diagonal is equal to $2R^-$ and edge to $2R^-/\sqrt{2}$ or $R^-\sqrt{2}$. Finally, the cube diagonal is equal to $R^-\sqrt{2}\sqrt{3}$, when the anion and cation are in contact:

$$R^+ + R^- = \sqrt{2}\sqrt{3}\, R^-/2 \text{ or } R^+ + R^- = \sqrt{3}\, R^-/\sqrt{2} \text{ or } R^+/R^- = \frac{\sqrt{3}}{\sqrt{2}} - 1 = 0.225$$

Coordination 4 applies starting with ratio R^+/R^- of 0.225 and until this ratio reaches the value for coordination 6, which is 0.414.

NOTE 3.3.– It is possible for an anion to be in coordination 4, at the center of a square plane formed of four cations (5), as in perovskite (SrTiO₃), in which the oxygen ion is surrounded by four strontium cations (see Figure 3.6).

But it is the octahedron formed by the cations surrounding the titanium ion (5) that defines the parameters of the unit cell and/or the cuboctahedron around the strontium ion, which can be visualized in Figures 3.6 and 3.9.

3.2.2.2. Coordination number 6

In general, in six-coordinate structures (see Figure 3.7), six anions form an octahedron (2 and 3) at the center of which a cation such as NaCl (1 and 2) glides.

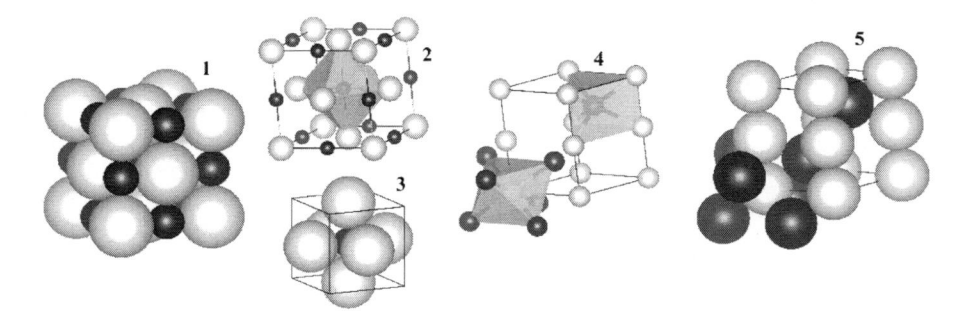

Figure 3.7. *Examples of six-coordinate structures (NaCl and NiAs). For a color version of this figure, see www.iste.co.uk/valls/inorganic.zip*

The four central anions (3) that form a square of side $2R^-$ are preserved, the square diagonal is equal to $2R^-\sqrt{2}$ and the semi-diagonal is $R^+ + R^-$.

It can be written as $R^+ + R^- = 2\sqrt{2}\,R^-/2 = R^-\sqrt{2}$ or $R^+/R^- = \sqrt{2} - 1 = 0.414$.

Coordination 6 is applicable starting with a steric criterion of 0.414 and until it reaches the value for coordination 8, which is 0.731.

NOTE 3.4.– Coordination number is equal to 6 for trigonal prism (4); these two types are present in nickel arsenide (NiAs) as $NiAs_6$ octahedron (4 and 5 in the lower part of Figure 3.7) and $AsNi_6$ trigonal prism (4 and 5 in the upper part of Figure 3.7). In this case, the bond is not ionic and Pauling's rules are no longer applicable.

Mixed six-coordinate complexes for anions (with two types of cations) are possible, for example, in perovskite ($SrTiO_3$), where oxygen is surrounded by a square plane of Sr^{2+} and two Ti^{4+} that form an octahedron in Figure 3.15.

3.2.2.3. *Coordination number 8*

In general, in eight-coordinate structures, eight anions form a cube (see Figure 3.8) at the center of which a cation (1) glides, as in CsCl (1 and 2).

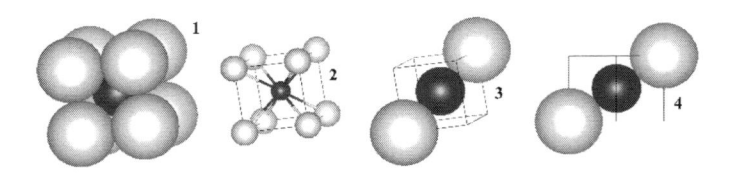

Figure 3.8. *Example of eight-coordinate structure (CsCl)*

The same cube formed of eight anions in contact (3 and 4) is preserved; therefore, $a = 2R^-$.

The cube diagonal is equal to $a\sqrt{3}$; therefore, $2R^-\sqrt{3} = 2(R^+ + R^-)$ or $\sqrt{3} = R^+/R^- + 1$ and $R^+/R^- = \sqrt{3} - 1 = 0.732$

NOTE 3.5.– Eight-coordinate structures in the form of anti-prisms with square base $NbTe_8$ can be observed in systems of $NbTe_4$ type [TAD 90] but these are MX_4 compounds, which are not covered in this book.

Coordination 8 is applicable starting with a steric criterion of 0.732 and until it reaches the value for coordination 12, which is 1.

3.2.2.4. Coordination 12

In general, in 12-coordinate structures, 12 anions form a cuboctahedron (see Figure 3.9) at the center of which a cation (1) glides. An example is $SrTiO_3$, where strontium ion is surrounded by 12 oxygen ions (1, 2 and 3).

No calculation is needed for coordination 12, as the anion and cation have identical sizes or $R^+/R^- = 1$. It suffices to remember the packings in metals and the cuboctahedron formed in ABC packings.

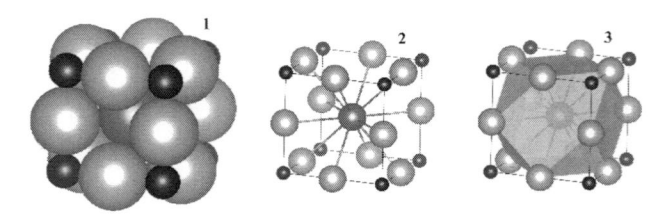

Figure 3.9. *Example of 12-coordinate structure ($SrTiO_3$). For a color version of this figure, see www.iste.co.uk/valls/inorganic.zip*

3.2.2.5. *Coordinations 2 and 3*

Coordinations 2 and 3 are proposed after the most common coordinations (4, 6, 8 and 12) as they are rare and bring little information, all the more so as they form no polyhedron but a triangle or a triangular base pyramid and a three-ion alignment.

3.2.2.5.1. Coordination 3

In general, in three-coordinate structures, three anions form an equilateral triangle at the center of which a cation glides (see Figure 3.10).

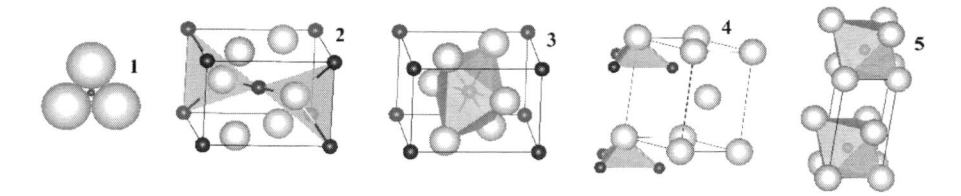

Figure 3.10. *Example of three-coordinate structure (TiO_2). For a color version of this figure, see www.iste.co.uk/valls/inorganic.zip*

NOTE 3.6.– As far as anions are concerned, three-coordinate structures can be observed, for example, for oxygen (3), in rutile (TiO_2) (see Figure 3.10). More specific coordinations can also be observed for the chlorine anion, which is at the top of a triangular base pyramid formed of three cations (5) for $CdCl_2$.

In both cases, it is the octahedral coordination of cations (3 and 5) that sets the parameters of the unit cell.

When an equilateral triangle is formed of anions surrounding a cation, the side of the triangle is equal to $2R^-$ and the altitude to $2R^- \sin 60°$ or $R^-\sqrt{3}$. Knowing that in an equilateral triangle altitudes, center lines, bisectors and perpendicular bisectors coincide, the following expression can be written:

$$\frac{2}{3} R^-\sqrt{3} = 2R^-/\sqrt{3} = R^+ + R^-$$

Hence, dividing by R^-, $2/\sqrt{3} = R^+/R^- + 1$ or $R^+/R^- = \frac{2}{\sqrt{3}} - 1 = 0.155$

Coordination 3 applies starting with the ratio R^+/R^- being equal to 0.115 and until it reaches the value for coordination 4, which is 0.225.

3.2.2.5.2. Coordination 2

For coordination 2, no ratio R^+/R^- can be proposed, one ion being surrounded by two ions of opposite signs (see Figure 3.11) and no relation can be proposed, since there is no polyhedron. For example, for Cu_2O (1 and 2), the cation has coordination 2 with $R^+/R^- = 0.33$ and while these values do not bring specific information, the unit cell parameter can be deduced, as the semi-diagonal of the cube should be equal to $2(R^+ + R^-)$.

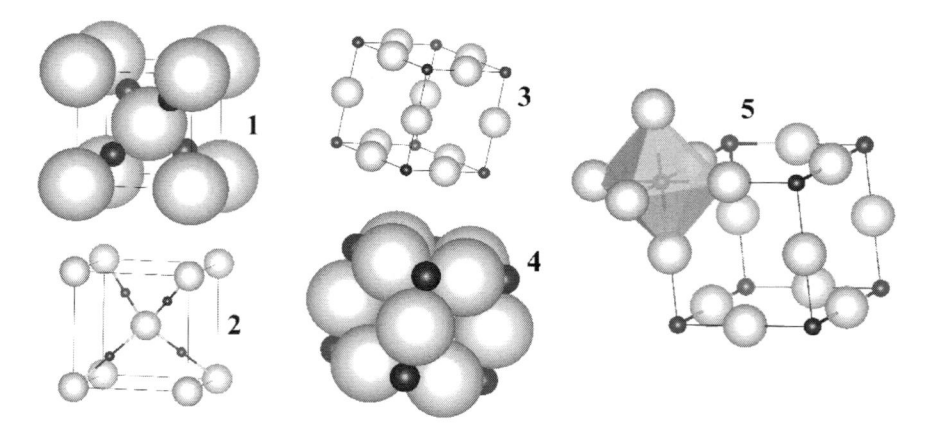

Figure 3.11. Examples of two-coordinate structures (Cu_2O and ReO_3). For a color version of this figure, see www.iste.co.uk/valls/inorganic.zip

NOTE 3.7.– For a complete description, coordination 2 for anions could be discussed, although it is rarely mentioned because it does not fix the geometry of the coordination polyhedron (Pauling's rules refer exclusively to cations).

For example, for ReO_3 (see Figure 3.11), coordination 2 can be observed (3 and 4) for oxygen anions, but it is the ReO_6 octahedron (5) that fixes the parameters of the unit cell.

The ions having this coordination form strongly orientated and mainly covalent bonds.

3.2.2.6. Coordination review

The values obtained are summarized in Table 3.2, which provides several examples of typical structures. The TiO_2 anion is mentioned, as it has been previously described, and the typical example of coordination 3, which is BN, is strongly covalent and does not satisfy the geometrical criteria described for this coordination.

The steric criterion should be used simply as an aiding tool, not as a means to justify the structures.

Indeed, of the 327 structures mentioned in this book, 233 are concerned by Pauling's rules and 53% confirm the expected steric criterion. This steric criterion is below the theoretical value for 31% of the compounds and above it for 16%.

R^+/R^-	Polyhedron	Coordination	Typical structure
–	Linear	2	Cu_2O
0.155 to 0.225	Triangle	3	TiO_2 (for the anion)
0.225 to 0.414	Tetrahedron	4	ZnS, SiO_2, Li_2O
0.414 to 0.732	Octahedron	6	$NaCl$, $NiAs$, TiO_2, $CdCl_2$, CdI_2, ReO_3
0.732 to 1	Cube	8	$CsCl$
1	Cuboctahedron	12	$SrTiO_3$ (for Sr^{2+})

Table 3.2. *Value of the steric criterion (R^+/R^-) for various coordination numbers*

The evolution of the steric criterion as a function of ionicity is represented in Figure 3.12 by colored points and the non-corresponding values are indicated by non-colored circles.

Only half of structures correspond to Pauling's first rule (verified on the basis of the steric criterion), which may lead to qualifying it as useless. But this is not the case because, if the rule is applicable, it relates to ionic compounds and, if not, the analysis of criteria can be conducted from a different perspective and bond-related information can be gathered.

Having this double objective in mind when observing crystals, this rule becomes much more effective than it may have seemed and allows for the analysis of all the ionic covalent binary compounds.

An approach to the interpretation of various criteria should clarify their limits, as electronegativity is used without adaptation, which impacts the value of ionicity.

Ionic radii are proposed depending on their coordination number without being modified as a function of their environment, which impacts the steric criterion.

This simplification is voluntary and, despite these approximations, it allows for simultaneous handling of several criteria without proceeding on a case-by-case basis, understanding the type of bond in the solids studied and making the best use of Pauling's rules.

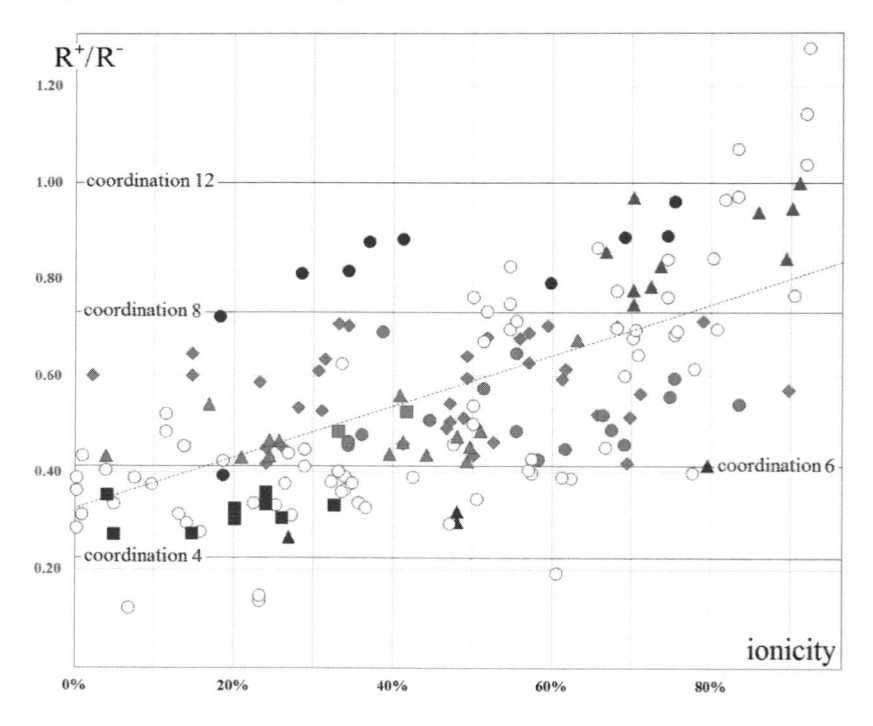

Figure 3.12. *Correlation between coordination and size of ions. For a color version of this figure, see www.iste.co.uk/valls/inorganic.zip*

NOTE 3.8.– The size of ions plays a role, but not in a simple manner, as other criteria must be simultaneously met.

In a reasoning uniquely based on the steric criterion (R^+/R^- = 0.404), a coordination 4 can be proposed, for example, for the CrO2 compound, but this is of rutile type, therefore coordination 6.

Coordination 4 is only present in ZnS and Li2O types, both being incompatible with the stoichiometry of the CrO2 formula.

The CrO2 compound adopts the rutile type that meets the stoichiometry and charge criteria (Pauling's second rule); therefore, the steric criterion plays no role.

Ionicity (57%) follows an ionic covalent bond, which is confirmed by an almost perfect availability (−0.1 pm) and indicates ion contact.

3.2.3. Pauling's second rule: electrostatic valence principle

Pauling's second rule states that the crystal should be overall neutral and hence contain the same number of positive and negative charges. However, nature is always energy-efficient and neutrality is confined to the smallest possible volume, which leads to the notion of coordination polyhedron formed of anions surrounding a cation.

Pauling defined the force of an ionic bond between M and X, when M^{m+} and X^{n-} ions are involved and the cation is surrounded by p anions:

force of M–X bond = charge of cation/number of neighboring anions = m/p.

An ion surrounded by p equivalent neighbors is linked to each of them and its charge is equally distributed among its neighbors. The sum of bonding forces for an ion should lead to the charge of the opposite ion.

For example, the bonding force of the titanium ion in TiO_2 is equal to 4 charges/6 neighbors = $\frac{2}{3}$ or 0.66.

Knowing that the oxygen ion is surrounded by three cations, 0.66 × 3 yields 2, which is the value of the charge of the oxygen ion (see Figure 3.13).

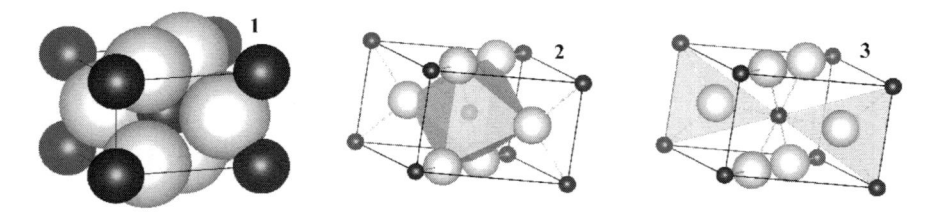

Figure 3.13. *Unit cell of TiO$_2$ with cationic and anionic coordination polyhedra. For a color version of this figure, see www.iste.co.uk/valls/inorganic.zip*

This type of calculation can be reversed in order to obtain coordination 3 of the oxygen ion (3), which is not readily visualized in a unit cell (1) like the one of the titanium ion (2).

A further type of calculation can be proposed, stemming from this bonding force, which is the rule of the electrostatic valence.

For the M_aX_b compound, it can be verified that:

a × coordination of M = b × coordination of X.

The above relation written, for example, for Cd_1Cl_2 yields $1 \times 6 = \underline{2} \times 3$.

While octahedral coordination 6 is easy to visualize in $CdCl_2$ (1) or CdI_2 (3), coordination 3 requires an effort (2 and 4), as the anion is not at the center of a polyhedron but constitutes one of the vertices (see Figure 3.14).

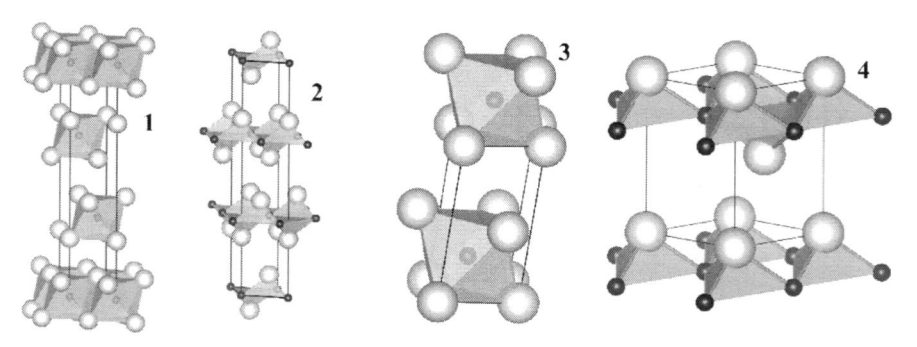

Figure 3.14. *Unit cells of CdCl$_2$ and CdI$_2$ with cationic and anionic coordination polyhedra. For a color version of this figure, see www.iste.co.uk/valls/inorganic.zip*

This rule explains a structure, such as perovskite, of formula SrTiO$_3$ (see Figure 3.15).

The strontium ion (2) in coordination 12 has a Sr–O bonding force of 2/12 or 0.17, and the titanium ion (1), in coordination 6 by O^{2-}, has a Ti–O bonding force of 4/6 or 0.67.

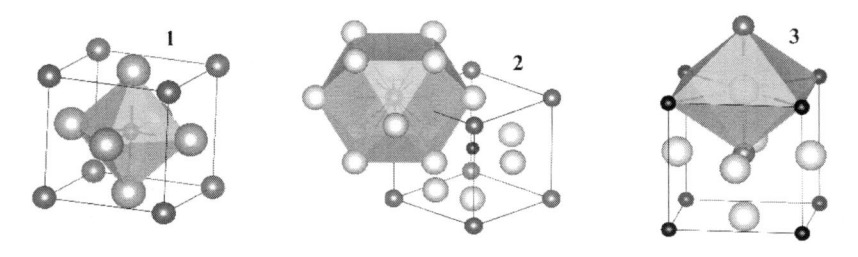

Figure 3.15. *Various coordination polyhedra in the SrTiO$_3$ unit cell. For a color version of this figure, see www.iste.co.uk/valls/inorganic.zip*

The only combination for the oxygen ion is 4 × 0.17 + 2 × 0.67 in order to obtain 2, which corresponds to its charge.

Every oxygen ion is therefore surrounded by four calcium ions and two titanium ions (3).

This information allows for full definition of the structural arrangement of perovskite of formula SrTiO$_3$ (see Figure 3.15).

The stoichiometry of the formula is therefore decisive for the crystal structure and, since the steric criterion has been associated with Pauling's first rule, formula stoichiometry is associated with the second rule.

NOTE 3.9.– The expression formula stoichiometry denotes a fraction that employs the indices of ions in the formula, for example, 1/1 for NaCl. It becomes 1/2 for rutile (TiO$_2$) and 2/1 for anti-fluorite (Li$_2$O).

Formula stoichiometry is useful when describing the crystalline change of a compound, as it can become a crippling criterion. This criterion is linked to Pauling's second rule, as stoichiometry sets the coordination of ions.

3.2.4. Pauling's third rule: connections of polyhedra

Pauling's third rule states that an anion coordination polyhedra is formed around each cation (and vice versa) and it is stable if the cation is in contact with each of its neighbors.

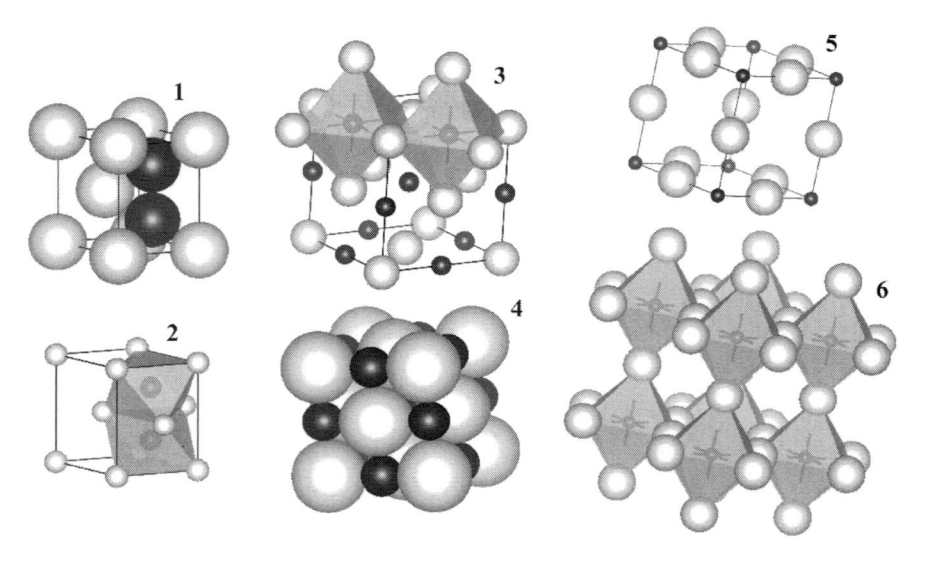

Figure 3.16. *Connection of coordination polyhedra for NiAs, NaCl and ReO$_3$. For a color version of this figure, see www.iste.co.uk/valls/inorganic.zip*

Coulomb interactions generate attraction forces that tend to bring cations and anions in contact, and, conversely, repulsive forces that act to separate cations.

Ionic crystals can then be considered assemblies of polyhedra connected through the faces, edges or vertices, which set the distance between the cations situated at the center of polyhedra (see Tables 3.3 and 3.4).

Crystals that have octahedra such as NiAs (1 and 2), NaCl (3 and 4) and ReO$_3$ (5 and 6) are represented in Figure 3.16, in which several octahedra have been chosen to visualize the connections.

Structure	Polyhedra	Connection	Distance between the centers of polyhedra
NiAs	octahedron	by faces (1 and 2)	$d_{Ni-Ni} = 0.7$ side of the octahedron (d_{As-As})
NaCl	octahedron	by edges (3 and 4)	$d_{Na-Na} = d_{Cl-Cl} = 1$ side of the octahedron
ReO$_3$	octahedron	by vertices (5 and 6)	$d_{Re-Re} = 1.41$ side of the octahedron (d_{O-O})

Table 3.3. *Distances between cations as function of the side of octahedra*

Crystals that have tetrahedra, such as CaF_2 (1 and 2), Cu_2O (3 and 4) and ZnS (5 and 6), are represented in Figure 3.17, in which several tetrahedra have been chosen to visualize the connections and relative positioning.

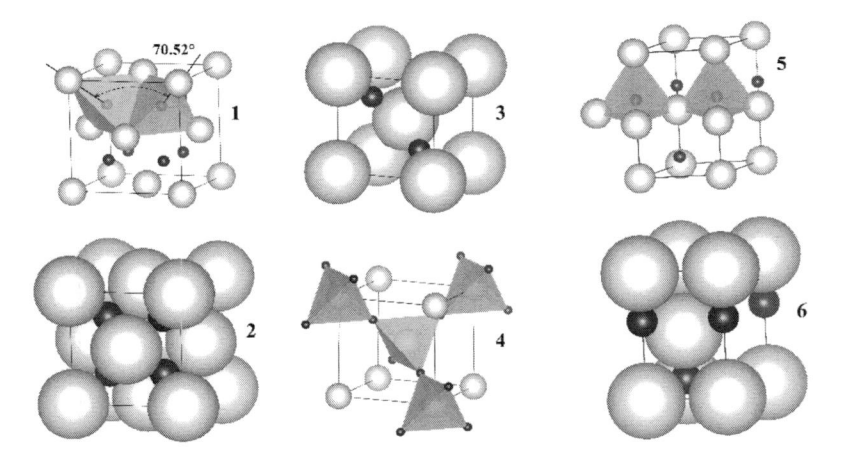

Figure 3.17. *Connection of coordination polyhedra for CaF_2, CuO_2 and ZnS wurtzite. For a color version of this figure, see www.iste.co.uk/valls/inorganic.zip*

It can be noted that sharing edges or faces moves the cations that are at the center of each polyhedron closer and increases electrostatic repulsions. It can be thought that identically charged ions are distributed so that the electrostatic energy is minimized. This rule is strictly followed by highly polar compounds such as fluorides and oxides but not by less polar compounds and it can even be contradicted by certain compounds.

Structure	Polyhedron	Connection	Distance between the centers of polyhedra
Li_2O	tetrahedron	by the side (1 and 2)	$d_{Li-Li} = 0.71$ side of the octahedron (d_{O-O})
ZnS	tetrahedron	by vertices (3 and 4)	$d_{Zn-Zn} = 1$ side of the tetrahedron (d_{S-S})
SiO_2	tetrahedron	by vertices (5 and 6)	$d_{Si-Si} = 1.22$ side of the tetrahedron (d_{O-O})

Table 3.4. *Distances between cations depending on the side of tetrahedra*

This rule may be in conflict with other criteria, for example, in the case of NiAs, in which face-sharing facilitates Ni–Ni bonds and confers it metallic character. It should be used with caution as, in many cases, it is a secondary criterion.

3.2.5. Pauling's fourth rule: separation of cations

Pauling's fourth rule states that in a crystal containing various cations, those of strong valence and low coordination do not share their polyhedron with the others.

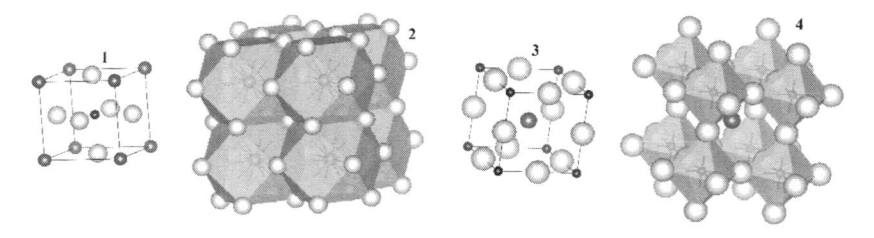

Figure 3.18. *Connection of coordination polyhedra for $SrTiO_3$. For a color version of this figure, see www.iste.co.uk/valls/inorganic.zip*

A good example is perovskite of formula $SrTiO_3$ (1 and 3), as in coordination 12: the strontium ion is surrounded by a SrO_{12} cuboctahedron that shares faces (2) with its neighbors, which does not move away the weakly charged cations. By contrast, the titanium ion is in coordination 6 and the TiO_6 (4) octahedra share only the vertices (see Figure 3.18), which moves the strongly charged titanium ions as far as possible.

This rule corresponds to very simple and logical considerations (repulsion between the ions of the same sign), but like the previous one, it involves a criterion that is often secondary and will not be always followed.

3.2.6. Pauling's fifth rule: homogeneity of the environment

Pauling's fifth rule states that the number of components of various types in a crystal tends to be small. To the extent that it is possible, chemically identical atoms have identical environments.

This rule, similar to the previous two, does not involve a main criterion and will often be disregarded.

> NOTE 3.10.– Crystal properties should be analyzed as everything else nature proposes and not as a simple equation with a unique solution. A set of simultaneously manifest interactions should be considered, one or several of which are dominant.

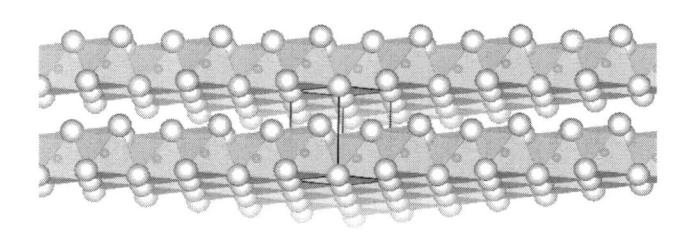

Pauling's rules are always used and, when they are not followed, a new criterion is manifest, such as the hydrogen bond (NH_4F), the metal–metal bond (NiAs) and a non-bonding electron pair (PbO). Then, the structures are less ionic, the increase in polarization in the bonding generates small dimensionality and layered structures can be observed ($CdCl_2$, CdI_2, etc.), such as CdI_2, as shown in the figure.

3.2.7. Presentation of criteria employed

In a first approach, five predominant interactions can be proposed and these will determine the crystal structure. The difficulty does not stem from the number of interactions considered but from their interdependency, which can be visualized in Figure 3.19 by arrows. Indeed, many others could be proposed (distance between cations, hydrogen bonding, etc.), but these generally relate to specific cases that will be mentioned as applicable.

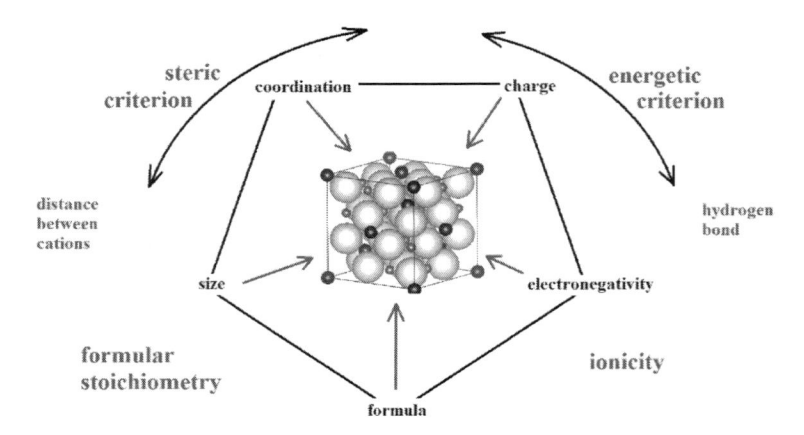

Figure 3.19. *Interactions defining crystal structure. For a color version of this figure, see www.iste.co.uk/valls/inorganic.zip*

Understanding a structure involves highlighting important interactions knowing that in a series there will always be specific cases, in which a minor interaction can no longer be ignored or becomes predominant for that compound.

In general, the most important interactions in a crystal are evaluated on the basis of the steric criterion and ionicity but these are completed in certain cases by the formula stoichiometry that does not allow for certain specific structures and the energetic criterion that will be identified, without being calculated, on the basis of its effects and the presence or absence of certain structures.

It should be noted that the charge and coordination of an ion define its size which, together with the size of anions and cations, allow for a definition of the steric criterion. Electronegativity (linked to ion valence) allows for a definition of ionicity. The chemical formula gives the stoichiometry, which will impose certain crystallographic types. An interdependency of criteria can be noted and this requires their simultaneous reading in order to understand the crystal.

It should be kept in mind that chemistry is an experimental science and observation of crystals is essential when drawing conclusions on a structure based on transparency, fragility, solubility, conductivity, melting temperature, etc. For example, the emergence of conducting properties in a crystal confirms the bond evolution toward metallic character or the loss of transparency signals an evolution toward a more covalent bond.

3.3. Geometry of binary crystals of MXn type

3.3.1. *Presentation of the mentioned compounds*

In Chapter 2, we described metallic crystals that contain only one type of atom linked by metallic bond. Ionic crystals contain ions and therefore at least one type of anion and one type of cation, bonded by an ionic or partially ionic bond.

In order to better highlight the geometry of structures, they are classified in ascending order of number of atoms in their formula:

– compounds of formula MX (CsCl of space group P m-3m, NaCl of group F m-3m, ZnS sphalerite of group F-43m or wurtzite of group P 6_3mc, NiAs of group P 6_3/mmc);

– compounds of formula MX_2 or M_2X (CaF_2, Li_2O of group F m-3m, TiO_2 rutile of group P 4_2/mnm, CdI_2 of group P-3m1, $CdCl_2$ of group R-3m);

– compounds of formula ABX_3 ($SrTiO_2$ mainly of group P m-3m) or AB_2X_4 ($MgAl_2O_4$ of group F d-3m).

It will be seen that while symmetry and neutrality limit the packing possibilities, nature combines the atoms in such a manner that the same empirical formula can cover very different structures. The only justification of this classification is to show the existence of several combinations for the same formula despite all the constraints of the ionic crystal.

In order to describe a structure, the lattice of each ion is described and the cations are considered to be located in the free spaces between anions around which coordinating polyhedra are formed (this image facilitates the three-dimensional view of crystals). Examples of the most varied possible compounds will be provided for each type, without approaching complexity.

The compounds that this book relies on (see Table 3.5) are binary compounds of MX and MX_2 type (with several examples of intermetallic compounds, when available) and ternary compounds of ABX_3 and AB_2X_4 type. Only the simplest compounds have been retained, the stoichiometric ones with one or two elements for the cations and one for anions, which limits their number. Several more complex compounds that belong to other space groups than the 10 presented will be simply mentioned for illustration purposes.

Compounds type	<	R^+ / R^-	>	Intermetallic	Non-ionic	Total
CsCl	8	12	–	–	9	29
NaCl	6	29	10	–	8	53
ZnS (s)	3	9	6	–	12	30
ZnS (w)	1	6	4	–	5	16
NiAs	–	–	–	4	22	26
CaF_2	10	9	–	8	–	27
Li_2O	–	4	10	4	–	18
CdI_2	18	15	–	–	–	33
$CdCl_2$	8	2	–	–	–	10
TiO_2	5	16	–	–	–	21

Table 3.5. *Number of compounds by structural type depending on the value of the steric criterion and on the bond type*

Models are used (perfect crystal and hard spheres) in order to calculate the criteria (ionicity, steric criterion and availability or modified availability) and to determine the dominant criterion in the structuring of the considered crystal. If certain criteria do not justify a structure, other criteria, which have become dominant, would be considered.

The compounds (see Table 3.5) are proposed by structural types, mentioning the value of the lower steric criterion (<), upper steric criterion (>) or in the expected interval (R^+/R^-) for the corresponding type of coordination. Certain compounds are not referred to, as they are non-ionic or intermetallic and a column has been dedicated to them. Details are provided on 19 perovskites and 29 spinels (37 and 27, respectively, which are simply mentioned), which leads to the presence of 305 compounds.

Of the total number of compounds, 77% are ionic covalent and the steric criterion will be applicable to them (for those 5% intermetallic compounds and 18% non-ionic compounds, the steric criterion is not applicable).

The hard sphere model will be applied to all the compounds in order to calculate availability (based on ionic radii) or modified availability (based on covalent or metallic radii). It should be remembered that, except for ternary compounds, 53% of these compounds have coordination according to the steric criterion, 31% above it and 16% below it.

3.3.2. *Study of cesium chloride*

3.3.2.1. *Description of the unit cell based on ion lattices*

The simplest structure of MX type is the ionic structure of CsCl, whose unit cell (1 and 2) has been represented. It features the first atom of the formula (Cs) at the cube vertex, while the following representations (3 and 4) feature the second atom of the formula (Cl) at the cube vertex. Although not a standard procedure, in order to simplify the descriptions in this book, the formulae are written so that the first atom is the one at the vertex of the unit cell (see Figure 3.20).

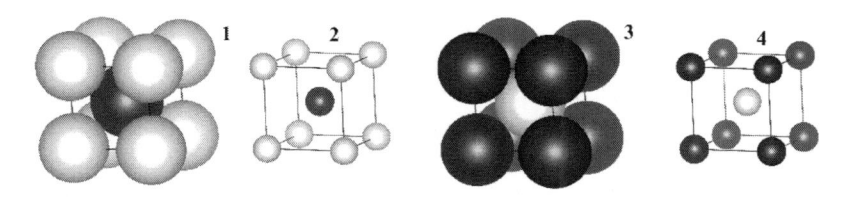

Figure 3.20. *Representation of CsCl unit cells (1 and 2) and ClCs unit cells (3 and 4)*

Thanks to these representations, a cubic P lattice formed by the cations (1 and 2) as well as an identical lattice formed by the anions (3 and 4) can be visualized. These two lattices are shifted by the semi-diagonal of the cube that corresponds to the distance between two ions of opposite charges.

Let us start from a CsCl (1) unit cell: adding the chlorine ions (see Figure 3.21) at the center of seven neighboring unit cells (2) and removing the cesium ions from the second unit cell yield the ClCs (3) unit cell.

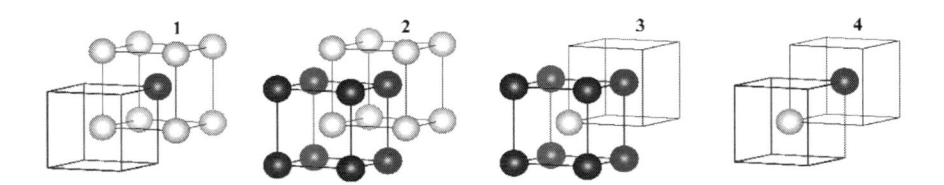

Figure 3.21. *Visualization of the pattern of CsCl unit cell*

Considering only the ions belonging to both unit cells, the number of atoms present in the unit cell can be obtained (4). It is the number of formula units of this unit cell, which is designated by Z (in this case, Z is equal to 1) and these two atoms are the pattern of the unit cell.

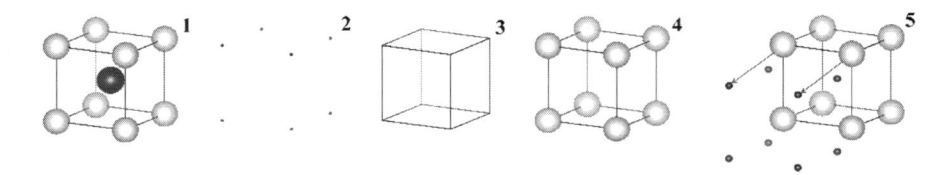

Figure 3.22. *Structuring of the CsCl unit cell*

The CsCl (1) structure proposed in Figure 3.22 comprises a cubic P lattice (2) that forms a cubic P unit cell (3), Cs^+ ions (4) at each node of this lattice and a cubic P lattice of Cl^- ions shifted by the cube semi-diagonal (5), of which only one is located in the unit cell (at the center of the cube).

3.3.2.2. *Description of the unit cell based on coordination*

Each ion is surrounded by ions of opposite signs that form a coordination polyhedron. The polyhedron formed by the chlorine ions is a cube whose lattice parameter is 412.3 pm [WYC 63b] and contains at its center a cesium ion.

Similarly, the cesium ions are in a cubic polyhedron formed of chlorine ions. The ions are said to be topologically identical, or simply they can be switched without modifying the crystalline structure.

Describing a crystal based on coordination polyhedra leads to using Pauling's rules and if the steric criterion is calculated (174 pm/181 pm [HAY 15]), the result is 0.96, which shows coordination 8 and leads to a cubic polyhedron.

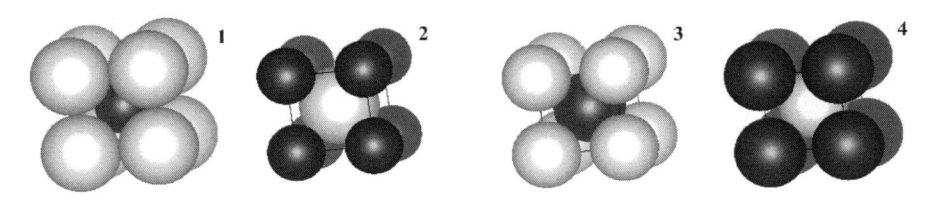

Figure 3.23. *Representation of ideal eight-coordinate ClCs and ClCs*

It should be noted that there is no contact between the ions (see Figure 3.23) and that only the ions of opposite sign are in contact along the diagonal of the cube (3 and 4). Indeed, the ratio is above 0.732 (see section 3.2.2), the limit value for which a maximum number of ions is in contact.

In the ideal structure of coordination 8 (1 and 2), the steric criterion is 0.732 and the anions are in contact (1), but not the cations (2).

3.3.2.3. *Calculation of various values of the unit cell*

Electronegativity is 0.79 for cesium and 3.16 for chlorine [HAY 15] and the bond is strongly ionic (theoretical ionicity is 75%).

Ionic radii [HAY 15] are denoted by $R^{+(VIII)}$ (174 pm) for the cation in coordination 8 and by $R^{-(VIII)}$ (181 pm) for the anion in coordination 8. If the size of the anion is considered fixed, the space available for the cation in the crystal can be calculated by deducting from the semi-diagonal of the cube ($a\sqrt{3}/2 = 357$ pm) the value 181 pm, which yields 176 pm. Therefore, the cation has 176 pm available for a proposed size of 174 pm, which proves that the hard sphere model is a perfect description of the crystal within 1%.

NOTE 3.11.– For a simple and rapid evaluation of the space available for cations, space availability is defined, which will be subsequently represented by the term availability.

It corresponds to the theoretical radius of the sphere in contact with all the anions of the coordination polyhedron from which anion radius is subtracted (in the previous example, it is +2 pm).

This availability should be assigned a tolerance and it has been arbitrarily fixed to 7.5 pm, which corresponds to the median of the values of the set of compounds presented. This choice allows for an overall interpretation of values without a case–by-case discussion and highlights the significant deviations.

Packing density is 0.669 if calculated with the values indicated for lattice parameter and ionic radii. It is below the theoretical value of 0.729 when all the ions are in contact in coordination 8, as the cations have moved away the anions by positioning in the cubic site (this is obviously just an image).

The calculation of d_{Cs-Cl} distances yields 357 pm, which is close to the sum of ionic radii (within 2 pm) and confirms the hard sphere model and a mainly ionic bond.

The e_{Cl-Cl} spaces (distances between the outer surface of two hard spheres) are 50.3 pm and e_{Cs-Cs} spaces are 64.3 pm, which show there is no contact between anions and between cations (see Figure 3.23).

The application of Pauling's rule of electrostatic valence to this compound M_aX_b makes it possible to verify that if cation coordination is 8, Pauling proposes $1(Cs^+) \times 8 = 1(Cl^-) \times 8$ for Cs_1Cl_1. It can be confirmed that anion coordination is also 8 and they are therefore topologically identical.

The structure can be described by the coordinates of cesium ions (0, 0, 0) and chlorine ions (½, ½, ½) at the center of the cube (or the inverse, since ions can be switched).

The type of lattice can be provided, cubic P in this case, and the point group P m-3m, which features all the symmetries of the simple cube, as the atom added to the center is the meeting point of all the elements of symmetry of this point group. Finally, the coordination of ions is given in the order cation and then anion, or 8:8 in this case.

The description of CsCl can be completed by justifying the number of formula units in the unit cell (Z), which is 1, as there are eight cations at eight vertices, which yields 1/8 in the unit cell, or one cation so that there is an anion at the center of the cube.

It can be highlighted (see Figure 3.24) that the elements of symmetry determine the dimensional parameters of the crystal and not vice versa. Indeed, the three axes of order 4 represented in the figure allow for the construction of the unit cell (2 and 3) based on the two atoms represented in the unit cell (1).

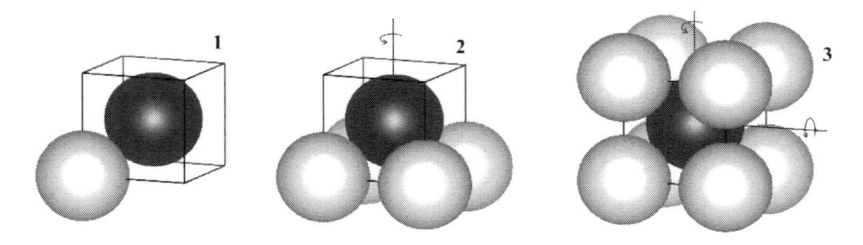

Figure 3.24. *Construction of the CsCl crystal based on a pattern and symmetries*

3.3.2.4. *Compounds of CsCl type*

The compounds of CsCl type are halides containing the biggest cations, such as Cs^+, Tl^+ and NH_4^+, and their number is limited compared to other types (for the ammonium ion, the size proposed by Pauling, highlighted below in italics, has been used).

An interpretation of the values provided in Table 3.6 can rely on the steric criterion, ionicity and availability. Indeed, if the model is adapted, the cation will have available at the center of the cube formed by the anions the space to get inserted and will be in coordination 8, therefore the steric criterion will be correct, as well as the availability. This is the case for most of them, and the hard sphere model is perfectly adapted for describing them. It can be proposed that the compounds involved are strongly ionic.

Compounds (MX)	R⁺; R⁻	a	Ionicity	R+/R⁻	Availability (pm)
NH₄Br [BAR 21]	*148*; 196	398.8	–	0.755	+1.4
NH₄Cl [VAI 56]	*148*; 181	386.0	–	0.818	+5.3
CsCl [WYC 63b]	174; 181	412.3	75%	0.961	+2.1
CsBr [BEC 25]	174; 196	428.6	69%	0.888	+1.2
CsI [NAT 29]	174; 220	456.7	60%	0.791	+1.5
TlCl [WYC 63b]	159; 181	383.4	37%	0.878	−8.0
TlBr [WYC 63b]	159; 196	397.0	29%	0.811	−11.2
TlI [WYC 63b]	159; 220	419.8	18%	0.723	−15.4

Table 3.6. *Several ionic covalent compounds of CsCl type*

The compounds are listed in the descending order of ionicity (except for ammonium salts, for which no electronegativity has been assigned to the ammonium ion) and it can be noted that availability becomes strongly negative when ionicity is below 30% (see Figure 3.25). This means that there is no more available space for the cation to be located in the cubic site, according to the hard sphere model.

This conclusion is obviously inappropriate since the crystal exists. It can therefore be proposed that covalence brings ions closer and reduces the distance between them; then, electron clouds interpenetrate and availability becomes negative.

The fact that the hard sphere model is applicable confirms the strongly ionic nature of bonds and the deviation from the model (strongly negative availability with correct steric criterion) shows the increase in covalence of the bonds.

The least-ionic compounds TlBr (29%) and TlI (18%) should have the highest part of covalence. Availability is negative (−11.2 and −15.4 pm, respectively) and confirms the part of covalence that determines, in the hard sphere model, the decrease in ion size. It can be noted that anionic availability is −0.2 pm for TlI, which is further confirmation of the needed reduction of ion size.

For compounds of CsCl type, structure formation may be mainly guided by the size of ions (Pauling's first rule) immediately followed by ionicity, since availability is correct.

When ionicity is below 30%, availability becomes negative (see Figure 3.25) and instead of invalidating models, the covalence part in the bond can be considered to become important for these compounds.

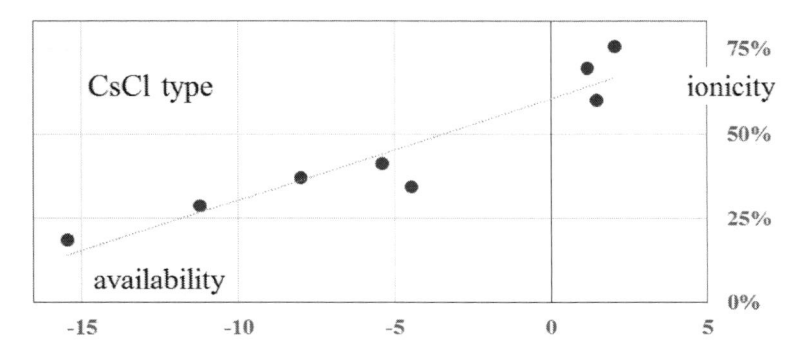

Figure 3.25. *Correlation between ionicity and availability for compounds of CsCl type*

We propose examples of intermetallic compounds that are of CsCl type but for which ionicity cannot be used, as the ionic bond cannot be considered.

Furthermore, the steric criterion makes no sense for these compounds, as there is no ionic radius to be taken into account.

On the contrary, the available place for atoms (not for ions) is always useful and therefore a first approximation involves the use of other types of radii.

Covalent radii and metallic radii allow for the calculation of a new availability that is qualified as modified availability that should be handled even more cautiously.

Indeed, crystals are only partially covalent or metallic and this will be shown here.

The use of this modified availability can indicate a tendency or certain consistency (see Table 3.7), as wider and wider approximations are made (metallic radii are given for a coordination that differs from the one they have in the compounds).

Compounds (MM')	R_M; $R_{M'}$	a	Modified availability (pm)
AuZn [WYC 63b]	130; $120/_{144;\ 133}$	319.0	$+26.3/_{-0.7}$
LiHg [WYC 63b]	130; $132/_{152;\ 151}$	328.7	$+22.7/_{-18.3}$
AgZn [WYC 63b]	136; $120/_{144;\ 133}$	315.6	$+17.3/_{-3.7}$
CuZn [WYC 63b]	122; $120/_{128;\ 133}$	294.5	$+13.0/_{-6.0}$
AuMg [WYC 63b]	130; $140/_{144;\ 160}$	325.9	$+12.2/_{-21.8}$
AgMg [WYC 63b]	136; $140/_{136;\ 140}$	330.2	$+10.0/_{-18.0}$
NiAl [WYC 63b]	124; $117/_{125;\ 143}$	288.1	$+8.5/_{-18.5}$
AgLi [WYC 63b]	130; $136/_{144;\ 152}$	316.8	$+8.4/_{-13.6}$
MgSr [WYC 63b]	140; $190/_{160;\ 215}$	390.0	$+7.7/_{-38.3}$
PdCu [WYC 63b]	122; $130/_{137;\ 128}$	298.8	$+6.8/_{-16.2}$
AgSc [ALD 62]	136; $159/_{136;\ 159}$	341.2	$+0.5/_{-10.5}$
AgLa [MET 77]	136; $194/_{145;\ 185}$	381.4	$+0.3/_{+0.7}$

Table 3.7. *Several intermetallic compounds of CsCl type*

Covalent radii lead to an availability of +15 pm on average above tolerance and metallic radii (indicated as indices) to an availability of −15 pm.

This inversion is linked to the increase in size of metallic radii compared to covalent radii. It can be thought that crystals are neither covalent (too high availability) nor metallic (too low availability) but partly covalent and partly metallic. On the contrary, there are no available criteria based on which one character or the other could be defined in terms of percentage.

The properties (conductibility, brittleness, aspect, etc.) allow for a more precise qualification of the type of bond for each compound.

For example, it can be noted that resistivity of alloys increases with compound proportion, reaches a maximum and then decreases.

Therefore, there is a loss of metallic character when the compound proportion increases, which involves a part of covalence (see, for example, the copper–aluminum alloy [HAY 15]).

> NOTE 3.12.– This approach is simple and consistent with the hypotheses on bond type. For example, if NiAl is represented with covalent radii (1), metallic radii (2) and Van der Waals radii (3), it can be noted that for covalent radii availability is positive (+8.5 pm), which involves a part of metallic character. Metallic radii lead to negative availability (−17.5 pm), which involves a part of covalence. Finally, Van der Waals radii yield an inconsistent result (−237.0 pm), which involves a bond between atoms, instead of packed atoms.

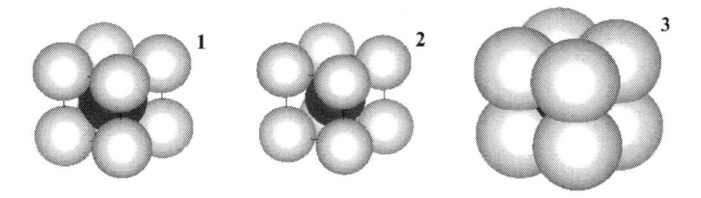

It can be thought that when covalent radii lead to strongly positive availability (AuZn) and metallic radii lead to a value within tolerance, the bond is more metallic. Conversely, for a strongly negative availability attached to metallic radii associated with a value within tolerance for covalent radii (MgSr), the bond is more covalent.

3.3.3. *Study of sodium chloride*

A further structure of MX type is that of sodium chloride (NaCl), with an empirical formula similar to CsCl but with a different space arrangement of the ions linked to the value of the steric criterion, which is smaller than that in the previous case. The conventional cubic F unit cell has been chosen (see Figure 3.26), and not the elementary rhombohedral unit cell, which does not allow for a visualization of all the crystal symmetries (see Figure 2.37).

3.3.3.1. *Description of the unit cell based on ion lattices*

The conventional unit cell of NaCl structure has been represented and the first ion of the formula (Na^+) can be observed (1) at eight vertices. In the ClNa (2) unit cell, the second ion, Cl^-, is at the vertex of the cube (see Figure 3.26).

It can be noted that cations and anions form similar cubic F packings (3 and 8) and hence they can be permuted, as in the case of CsCl.

Figure 3.26 shows a cubic F packing of sodium ions in the first unit cell (3) and a cubic F packing of chlorine ions in the last unit cell (8).

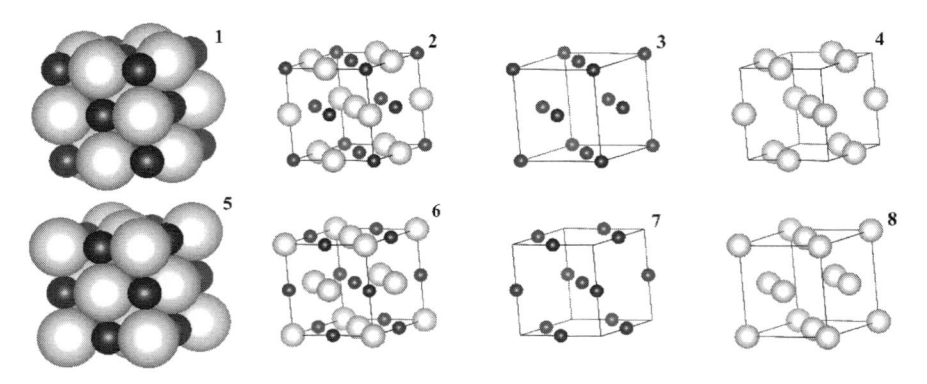

Figure 3.26. *Representation of conventional NaCl and ClNa unit cells*

In order to pass from one unit cell to the other, it suffices, for example (see Figure 3.27) starting from ClNa (1), to place a neighboring unit cell (2) and shift the unit cell by a half-lattice parameter (3) and to isolate the NaCl unit cell (4) at the center.

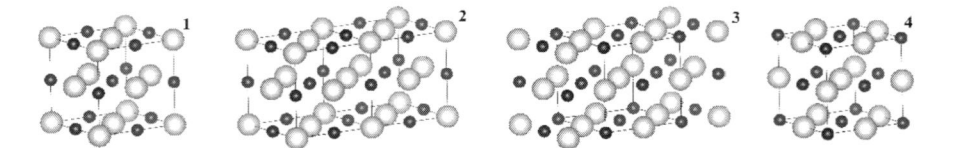

Figure 3.27. *Visualization of NaCl unit cell based on the ClNa unit cell*

The NaCl unit cell (1) is therefore composed (see figure 3.28) of a cubic F packing of Na$^+$ ions (2) and a cubic F packing of Cl$^-$ ions (4) shifted by the semi-edge of the cube corresponding to the unit cell (5).

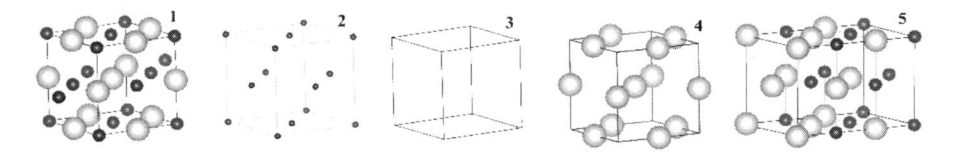

Figure 3.28. *Structuring of NaCl unit cell. For a color version of this figure, see www.iste.co.uk/valls/inorganic.zip*

Taking a unit cell as a starting point (1), an identical unit cell that is shifted by a quarter of a diagonal of the first unit cell (2) is added (see Figure 3.29).

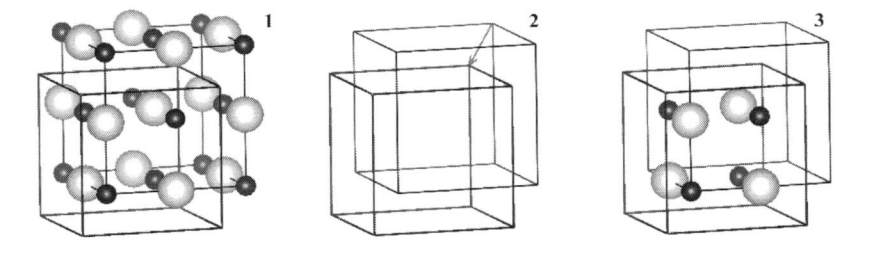

Figure 3.29. *Visualization of ions present in the NaCl unit cell*

If the representation only includes atoms that are shared by the two unit cells, the number of ions present in a unit cell (3) is visualized. This corresponds to four cations and four anions distributed alternatively on the eight vertices of a cube whose side is equal to half the lattice parameter.

Counting the fractions of each anion in the unit cell (1) yields the same result, as there are eight anions at the eight vertices that are 1/8 in the unit cell and hence 1 anion and six anions at the center of faces, which are 1/2 in the unit cell or three anions, therefore a total of 4. There are 12 cations at the center of edges (4), which are 1/4 in the unit cell or three anions, and one at the center of the unit cell, therefore a total of 4 ($Z = 4$).

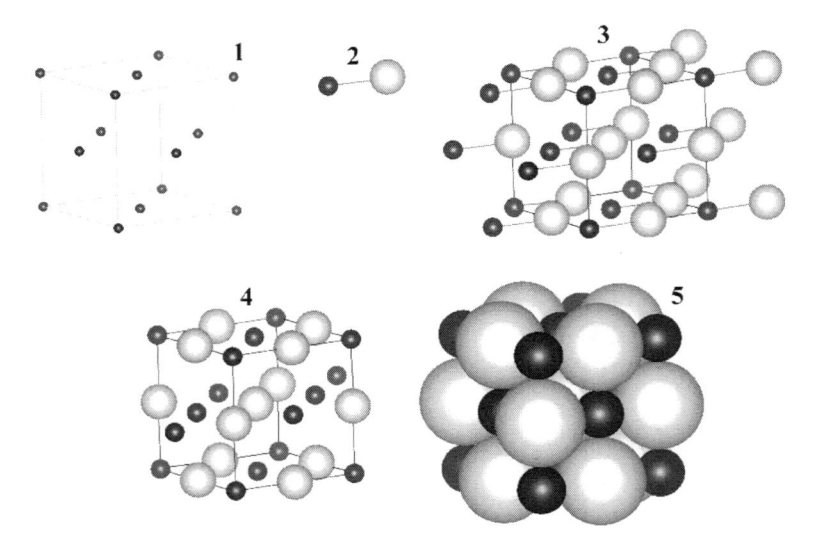

Figure 3.30. *Construction of the NaCl unit cell based on lattice and pattern. For a color version of this figure, see www.iste.co.uk/valls/inorganic.zip*

A crystal can be defined on the basis of its lattice (1) and pattern (2); by placing the pattern at each node, Figure 3.30 (3) is obtained, from which the NaCl unit cell can be readily extracted (4 and 5).

This is only an abstract description of the ion packing, but it has the advantage of exposing the brain to repeated three-dimensional visualization.

It is useful to conduct in parallel with the observation a software-assisted modeling of figures, involving their manipulation and choice of the clearest representation, conducted by a simple click. This type of visualization offers the possibility to rotate or enlarge a structure for better observation and to measure a distance or change a parameter and observe the effects.

3.3.3.2. *Description of the unit cell based on coordination polyhedra*

Since the ion packing is cubic F, the coordination polyhedra of anions or cations are cuboctahedra (see Figure 3.31) comprising only identical cations (1) or anions (3).

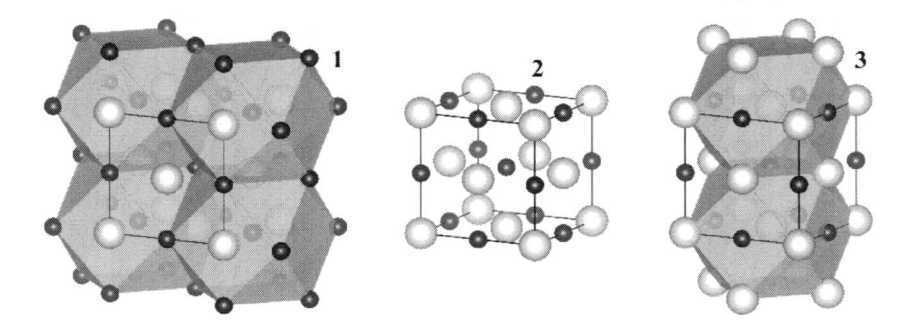

Figure 3.31. *Coordination polyhedron of identical ions in ClNa. For a color version of this figure, see www.iste.co.uk/valls/inorganic.zip*

These cuboctahedra are horizontally shifted by a half-edge of the unit cell (see Figure 3.32) and the previously described shift of cubic F packing can be noted.

The visualization of coordination polyhedra between opposite ions based on the ClNa unit cell (1 and 5) brings further information (see Figure 3.33).

The polyhedron around cations is an octahedron ($NaCl_6$) that contains one Na^+ ion at the center (2, 3 and 4), each edge of which has a length of $a / \sqrt{2}$. The steric criterion is equal to 0.55 and confirms a 6 coordination, therefore the octahedral polyhedron.

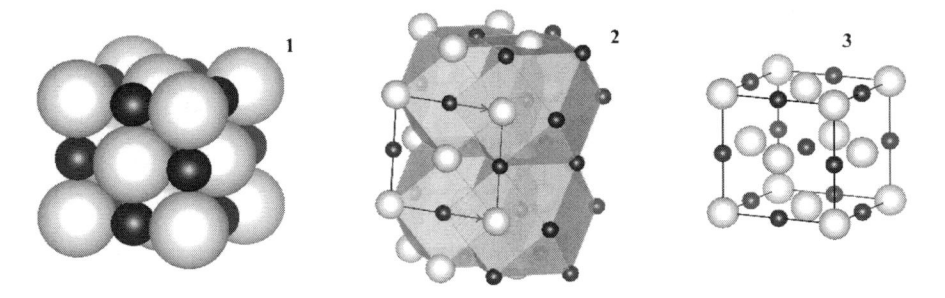

Figure 3.32. *Relative positioning of coordination polyhedra of identical ions in ClNa. For a color version of this figure, see www.iste.co.uk/valls/inorganic.zip*

Similarly, Cl^- ions are in the octahedral site ($ClNa_6$) formed by Na^+ ions (6, 7 and 8) as the ions can be switched without changing crystal structure (see Figure 3.33). In order to complete the coordination polyhedron, a cation external to the unit cell has been represented (5) (denoted by a cross).

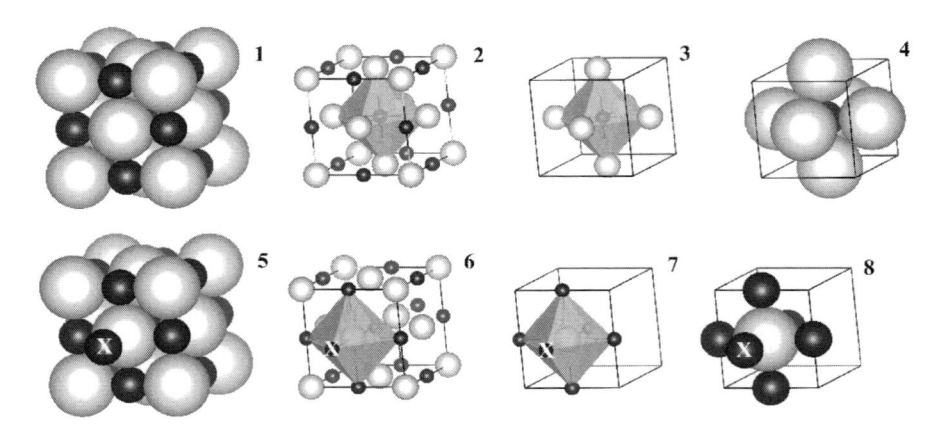

Figure 3.33. *Representation of $NaCl_6$ and $ClNa_6$ coordination polyhedra. For a color version of this figure, see www.iste.co.uk/valls/inorganic.zip*

It should be noted that there is no contact between the set of ions as in the ideal unit cell (1 and 2) and that only the ions of opposite sign are in contact (3 and 4) as the value of the steric criterion is above the limit value of 0.414 (see Figure 3.34).

The crystal can be described on the basis of coordination polyhedra ($NaCl_6$ or $ClNa_6$), and it can be noted they have a common edge. This is in agreement with Pauling's rules, as cations have few charges and they are all of the same type (see Figure 3.16).

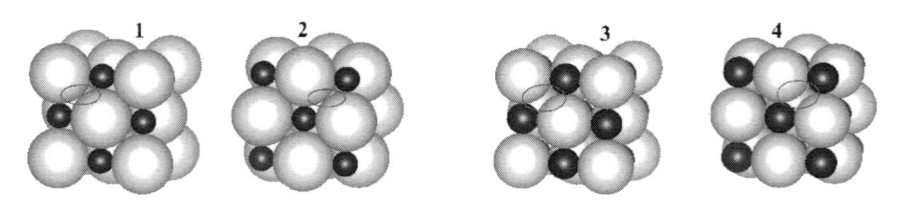

Figure 3.34. *Representation of ideal coordination 6 and of the ClNa and NaCl unit cell*

The visualization of polyhedra (2 and 3) is very useful but their value for crystal description purposes is much less so. The conventional unit cell (1 and 4) is preferred, which is more explicit on the overall set of crystal symmetries. Moreover, this volume can be readily memorized and manipulated (see Figure 3.35), for example in the form of molecular models.

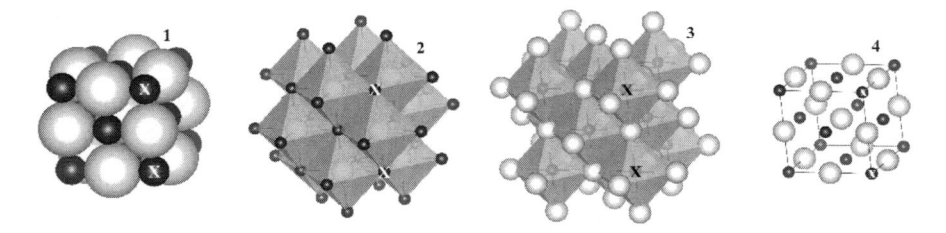

Figure 3.35. *Description of NaCl based on coordination polyhedra. For a color version of this figure, see www.iste.co.uk/valls/inorganic.zip*

In order to correctly situate $NaCl_6$ (2) and $ClNa_6$ (3) polyhedra, a cross has been used to signal two similar atoms in the unit cell in compact model (1) or unfolded model (4).

3.3.3.3. *Calculation of various values of the unit cell*

Electronegativities of sodium and chlorine are 0.93 and 3.16 [HAY 15], respectively, and the bond is strongly ionic (ionicity of 72%).

The ionic radius of cations ($R^{+(VI)}$) is 102 pm and that of anions ($R^{-(VI)}$) is 181 pm [HAY 15], with a lattice parameter a equal to 562.0 pm [WAL 04]. Availability is obtained by deducting 181 pm from the semi-edge, which gives 100 pm. The Na^+ cation has 100 pm available for a size of 102 pm, which is within 2%; the ion has available the needed space, therefore the hard sphere model perfectly describes the crystal.

The calculation of packing density yields 0.66 with the indicated values of the lattice parameter and the radii of ions far from the theoretical value of 0.793 when all the ions are in contact in coordination 6. The cation that is larger than the limit value displaces the anions and lowers the packing density.

The calculation of d_{Na-Cl} distances yields 281 pm, very close to the sum of ionic radii (283 pm) and confirms the mainly ionic bond. The value of e_{Cl-Cl} spaces is 35.4 pm and that of e_{Na-Na} is 193.4 pm, which shows the absence of contact between anions and between cations, according to the steric criterion value.

Finally, the anion–cation spaces e_{Na-Cl} (or availability) are of −2 pm and confirm the validity of models and the mainly ionic character of the bond.

Applying the electrostatic valence rule allows verifying that if cations have coordination 6, Pauling proposes it can be written 1 (Na^+) × 6 = 1 (Cl^-) × 6 for Na_1Cl_1, which confirms that ions are topologically identical and have the same coordination.

The structure can be described by the coordinates of Cl^- atoms (0, 0, 0) and (½, ½, 0) and those of Na^+ (½, 0, 0) and (½, ½, ½) or the inverse, since ions can be permuted.

Ions have cubic F packing of F m-3m point group.

The description can be completed by the number of formula units in the unit cell (Z), which is 4 since there are 8 × 1/8 (Cl^-) + 6 × 1/2 (Cl^-) or four anions and 12 × 1/4 (Na^+) + 1 (Na^+) or four cations and therefore Z is 4, and finally by the coordination of ions, which is 6:6.

The NaCl crystal takes the form of two F packings shifted by the semi-edge of the unit cell (see Figure 3.27) or of a cubic F packing whose octahedral sites are all filled by ions of opposite charge (see Figure 3.35).

3.3.3.4. Compounds of NaCl type

As there are many compounds of NaCl type, only the binary ones have been chosen, covering approximately 20 alkaline halides (except for CsCl, CsBr and CsI, which are of CsCl type); approximately 30 oxides, sulfides, selenides and tellurides and approximately 10 nitrides, carbides and hydrides (see Tables 3.6, 7 and 8). This list can be presented starting from cations, which are more than 20 alkaline, more than 10 alkaline-earth and more than 15 elements of d-block.

The compounds in Table 3.8 are given in descending order of ionicity and it can be seen that the steric criterion is below the expected minimum value, but availability is correct. The hard sphere model is therefore correct, although Pauling's first rule is not followed.

Compounds (MX)	R^+; R^-	a (c)	Ionicity	R^+/R^-	Availability (pm)
LiBr [OTT 23]	76; 196	550.1	62%	0.388	+3.1
LiI [OTT 23]	76; 220	600.0	51%	0.345	+4.0
ZrS [WYC 63b]	74; 184	525.0	34%	0.402	+6.5
MgSe [WYC 63b]	72; 198	545.1	34%	0.364	+2.6
MgS [WYC 63b]	72; 184	520.3	34%	0.391	+4.2
ZnTe [HOL 68]	74; 221	610.3	5%	0.335	+10.2
ZnTe type ZnS (s)	60; 221	610.2	5%	0.271	−16.8
ZnTe type ZnS (w)	60; 221	427 (699)	5%	0.271	−19.4

Table 3.8. *Several compounds of NaCl type (R^+/R^- < 0.414)*

In this case, the energetic criterion and the formula stoichiometry outweigh the steric criterion and the compounds do not change their coordination to 4.

Their only possibility would be to change into ZnS sphalerite or wurtzite, which not only have equivalent environments for the two ions but are also strongly covalent compounds, therefore incompatible with the value of ionicity in this series.

NOTE 3.13.– The only compound (ZnTe) that is beyond tolerance for the steric criterion and has very low ionicity exists in coordination 4 in the form of crystalline types ZnS sphalerite [HOL 68] and wurtzite [WYC 63b]. It is quite the exception that proves the rule and shows the part of covalence in the bond.

In Table 3.9, the compounds have a value of steric criterion above the maximal value of 0.732 for coordination 6. They are listed in descending order of the value of ionicity and it can be noted that the availability is always correct, despite the high ratio.

Compounds (MX)	R^+; R^-	a	Ionicity	R^+/R^-	Availability (pm)
CsF [WYC 63b]	170; 133	600.8	92%	1.278	−2.6
KF [WYC 63b]	138; 133	534.7	92%	1.038	−3.6
RbF [WYC 63b]	152; 133	564.0	92%	1.143	−3.0
NaF [WYC 63b]	102; 133	462.0	91%	0.767	−4.0
BaO [WYC 63b]	135; 140	552.3	82%	0.964	+1.1
SrO [VER 09]	118; 140	516.1	80%	0.843	+0.1
KCl [WYC 63a]	138; 181	628.8	75%	0.762	−4.6
RbCl [WYC 63b]	152; 181	658.1	75%	0.840	−3.9
RbBr [WYC 63b]	152; 196	685.5	68%	0.776	−5.3
AgF [WYC 63b]	115; 133	492.0	66%	0.865	−2.0
RbCl type CsCl	152; 181	374.0	75%	0.840	−9.1

Table 3.9. *Several compounds of NaCl type (R^+/R^- > 0.732)*

Maintaining coordination 6 for these compounds minimizes crystal energy, and the steric criterion becomes secondary. Ionicity plays no role, as it is significant for all the compounds and allows them to be perceived as mainly ionic and therefore the hard sphere model is also applicable in this case.

NOTE 3.14.– The RbCl compound, which is of NaCl type as well as CsCl type, can serve to illustrate the role of the energetic criterion without any calculation.

This compound is indeed stable when of NaCl type, but does not exist as CsCl type unless under high pressure [WYC 63b], which confirms that the energetic criterion outweighs the steric criterion.

Availability in NaCl type is within tolerable limits and it validates the hard sphere model, whereas in CsCl type, it is −9.1 pm (see Table 3.9) and therefore incompatible with the significant value of ionicity (75%).

This negative availability does not confirm a part of covalence but the fact that there is ion contraction under these extreme conditions, although this subject is beyond the scope of the present book.

Table 3.10 presents compounds whose steric criterion corresponds to coordination 6 and it can be noted that the overall availability validates the hard sphere model.

Several compounds with ionicity below 30% (MnSe, SrTe, AgBr and AgCl) have strongly negative tolerance, which does not invalidate the models but indicates a significant part of covalence in their bond.

NOTE 3.15.− Experimental parameters can be used, such as the melting temperature.

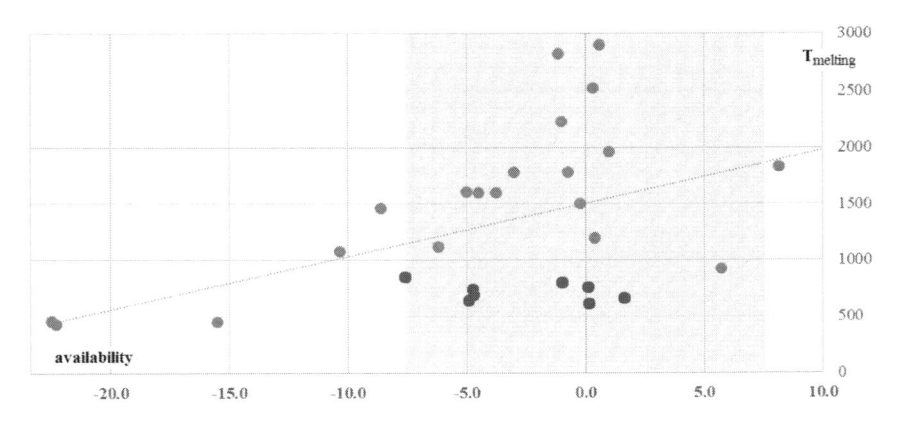

For a color version of this figure, see www.iste.co.uk/valls/inorganic.zip

The compounds with the most negative availability are those with the lowest melting temperature [HAY 15] except the alkaline and alkaline-earth derivatives indicated in different colors.

Compounds (MX)	R^+; R^-	A	Ionicity	R^+/R^-	Availability (pm)
LiF [OTT 26]	76; 133	402.8	90%	0.571	−7.6
CaO [OFT 27]	100; 140	481.2	79%	0.714	+0.6
NaCl [WAL 04]	102; 181	564.1	71%	0.564	−1.0
MgO [SAS 79]	72; 140	421.7	70%	0.514	−1.2
LiCl [OTT 23]	76; 181	514.3	70%	0.420	+0.1
KBr [OTT 26]	138; 196	658.5	68%	0.704	−4.8
NaBr [OTT 26]	102; 196	596.2	66%	0.520	+0.1
MnO [TAY 84]	83; 140	444.5	61%	0.593	−0.8
KI [WYC 63b]	138; 220	706.6	57%	0.627	−4.7
ZnO [BAT 62]	74; 140	428.0	57%	0.529	+0.0
RbI [WYC 63b]	152; 220	734.2	57%	0.691	−4.9
CdO [ZHA 99]	95; 140	469.5	56%	0.679	−0.2
NaI [WYC 63b]	102; 220	647.3	53%	0.464	+1.6
BaSe [WYC 63b]	135; 198	660.0	52%	0.682	−3.0
BaS [WYC 63a]	135; 184	638.8	52%	0.730	+0.4
SrS [WYC 63b]	118; 184	602.0	49%	0.641	−1.0
SrSe [WYC 63b]	118; 198	623.0	49%	0.596	−4.5
CoO [RED 62]	65; 140	424.0	48%	0.464	+7.0
NiO [AND 01]	69; 140	420.0	47%	0.493	+1.0
CaS [OFT 27]	100; 184	568.6	47%	0.543	+0.3
CaSe [OFT 27]	100; 198	591.2	47%	0.505	−2.4
AgCl [WYC 63b]	115; 181	547.0	31%	0.635	−22.5
BaTe [WYC 63b]	135; 221	698.6	31%	0.611	−6.7
SrTe [WYC 63b]	118; 221	647.0	28%	0.534	−15.5
CaTe [WYC 63b]	100; 221	634.5	26%	0.452	−3.8
MnS [OTT 26]	83; 184	524.0	24%	0.451	−5.0
MnSe [WYC 63b]	83; 198	544.8	24%	0.419	−8.6
AgBr [WYC 63b]	115; 196	577.5	23%	0.587	−22.3
PbS [WYC 63b]	119; 184	593.6	15%	0.647	−6.2
PbTe [WYC 63b]	119; 198	645.4	2%	0.601	−17.3

Table 3.10. *Several compounds of NaCl type (0.732 < R^+/R^- < 0.414)*

Similar to the other series, it can be noted that when ionicity is low, availability can become negative, which confirms a part of covalence (MnSe, SrTe, PbTe, AgBr and AgCl).

Table 3.11 lists the compounds with low ionicity and a strong part of covalence in their bond, whereas the steric criterion and availability are not applicable. The use of modified availability calculated on the basis of covalent radii is only an approximation but it shows a significant part of covalence in the bond.

Compounds (MX)	R; R'	a	Modified availability (pm)
CrN [NAS 77]	130; 71	414.8	+6.4
LaN [WYC 63b]	194; 71	530.0	+0.0
VN [BEC 25]	144; 71	430.0	+0.0
TiC [AIG 94]	148; 75	435.0	−5.5
TiN [BAN 41]	148; 71	424.4	−6.8
UN [WYC 63b]	183; 71	488.4	−9.8
UC [WYC 63b]	183; 75	495.9	−10.1
TaC [WYC 63b]	143; 75	445.4	−10.3

Table 3.11. *Several non-ionic compounds of NaCl type*

The limits of the proposed models have been reached but any chemist should be able to obtain more information, for example, by measurements of conductivity, which may help in proposing the parts of metallic and covalent character.

3.3.4. *Study of zinc sulfide (sphalerite)*

A new structure of MX type is that of ZnS (sphalerite), having the same empirical formula as CsCl or NaCl with a cubic unit cell. The spatial arrangement of ions is different, as the coordination of cations is 4 and each ion is surrounded by a tetrahedron of ions of opposite sign.

In this case, a conventional cubic F unit cell (1 and 2) has also been chosen, instead of the elementary rhombohedral unit cell (3), which is more difficult to handle (see Figure 3.36).

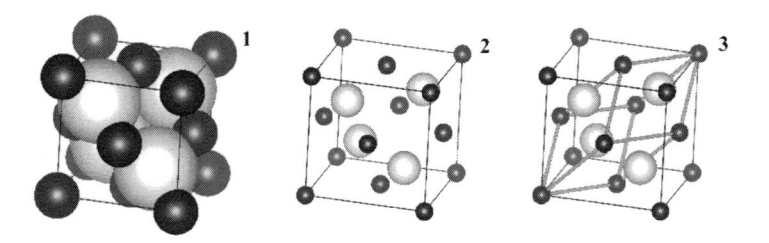

Figure 3.36. *Representation of the conventional and simple ZnS (sphalerite) unit cell*

3.3.4.1. Description of the ZnS (sphalerite) unit cell based on ion lattices

The packing of ions in ZnS is still cubic F but the distribution of ions differs from the previous ones and corresponds to a cubic F packing of ions S^{2-} in which a tetrahedron formed of Zn^{2+} ions (1) or its inverse (2) can be observed (see Figure 3.37).

In order to visualize the two SZn and ZnS packings, it is sufficient to place S^{2-} (1) and Zn^{2+} (2) alternatively at the vertex of the unit cell (see Figure 3.38) and it can be verified that ions are topologically identical.

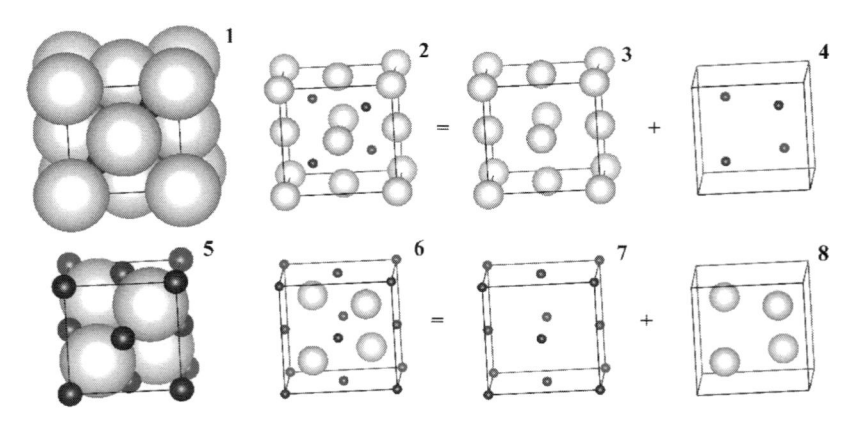

Figure 3.37. *Representation of SZn and ZnS (sphalerite) unit cell*

The number of formula units (Z) is the result of a simple counting of fractions of each cation in the unit cell as there are eight cations at eight vertices that are 1/8 in the unit cell and six cations at the center of the faces that are half in the unit cell, or 8 × 1/8 + 6 × 1/2 or four cations (see Figure 3.38). For anions, the calculation is simple as all four of them are entirely in the unit cell. Therefore, there are four formula units in this conventional unit cell and Z is equal to 4.

By adding several well-chosen atoms of neighboring unit cells (1), the two cubic F packings and the shift by a quarter of the diagonal of the unit cell (2 and 3) can be clearly seen and a two-atom pattern can be deduced (4).

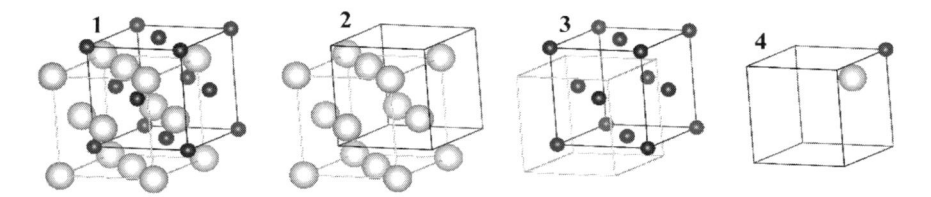

Figure 3.38. *Structuring of ZnS (sphalerite) unit cell*

3.3.4.2. *Description of ZnS (sphalerite) unit cell based on coordination polyhedra*

The coordination polyhedron of zinc ions (1) or sulfur ions (2) is a cuboctahedron, as cubic F packings are involved (see Figure 3.39).

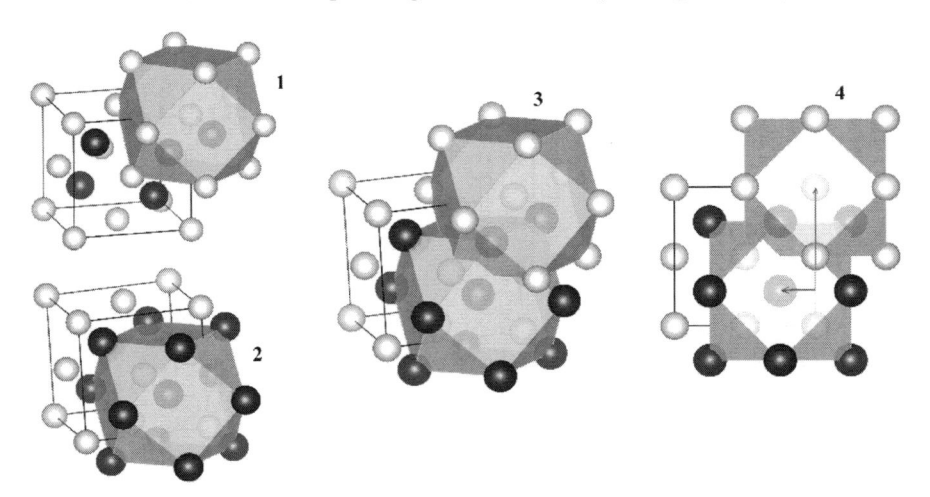

Figure 3.39. *Coordination polyhedron of identical ions in ZnS (sphalerite) type. For a color version of this figure, see www.iste.co.uk/valls/inorganic.zip*

Their relative positioning (3) has been indicated and a shift of the centers of the two cuboctahedra can be noted, of a/4 along one axis and of 3a/4 along the other axis (4).

The coordination polyhedron of ions, formed of sulfur ions around zinc ions, is a tetrahedron and contains a zinc ion at the center. The same tetrahedron is formed of zinc ions around sulfur ions.

These tetrahedra can be used to describe the unit cell (1 and 2) and if a ZnS_4 tetrahedron is placed on each of the zinc ions in the unit cell, then (3) is obtained. Several ions from outside the unit cell should be added in order to visualize SZn_4 tetrahedra (4) that correspond to a cubic F distribution of these tetrahedra (see Figure 3.40).

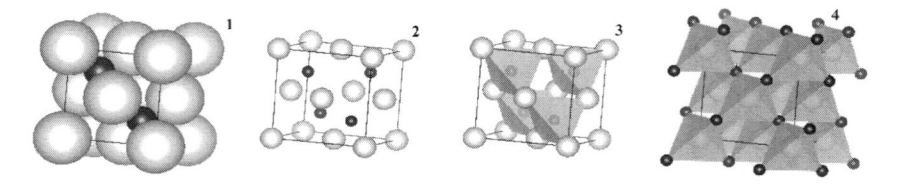

Figure 3.40. *Representation of ZnS_4 and SZn_4 coordination polyhedra. For a color version of this figure, see www.iste.co.uk/valls/inorganic.zip*

The inverse ZnS unit cell (1 and 2) allows for the visualization (see Figure 3.41) of SZn_4 tetrahedra (3) and, if several ions from outside the unit cell are added, a tetrahedra distribution identical to the one in the previous figure can be visualized (4).

The pattern is formed of two ions as in the case of CsCl (semi-diagonal of the unit cell) or of NaCl (semi-edge of the unit cell) but, in the case of ZnS, it is a quarter of the diagonal of the cube formed by the unit cell.

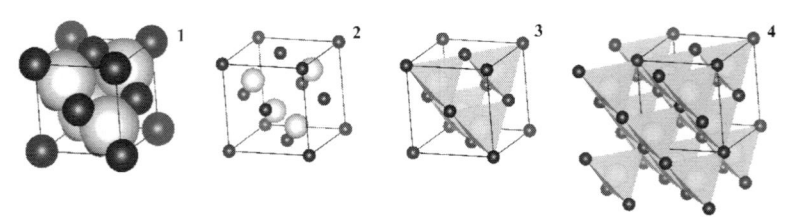

Figure 3.41. *Construction of the ZnS unit cell based on SZn_4 tetrahedra. For a color version of this figure, see www.iste.co.uk/valls/inorganic.zip*

3.3.4.3. Calculation of various values of the ZnS (sphalerite) unit cell

Electronegativities [HAY 15] for zinc and sulfur are 1.65 and 2.6, respectively, as the bond is weakly ionic (ionicity of 20%).

Ionic radii [HAY 15] $R^{+(IV)}$ and $R^{-(IV)}$ are equal to 60 and 184 pm, respectively, with the lattice parameter a being equal to 547.5 pm [YIM 69].

Availability is obtained by deducting from the quarter of the cube diagonal the ionic radius of the anion, which is 50.2 pm ($a\sqrt{3}$–184). The cation has available 50.2 pm for a size of 60 pm; therefore, the hard sphere model cannot be used for the description of this weakly ionic crystal.

The calculation of packing density leads to 0.68, which is far from the theoretical value of 0.748 when all ions are in contact (1) in coordination 4. The cation whose size exceeds the limit value has displaced the anions and consequently lowered the packing density.

The calculation of d_{Zn-S} distances yields 234.2 pm < $R^+ + R^-$ (10.8 pm) and confirms the weakly ionic bond (20%). The spaces e_{S-S} and e_{Zn-Zn} equal to 7.2 and 382.5 pm, respectively, show the absence of contact between the anions and between the cations. On the contrary, if the model shows covalence, there is no simple relation between the percentage of ionic character and the size of ions.

The application of the electrostatic valence rule for the M_aX_b compound makes it possible to verify that if the coordination of cations is 4, for Zn_1S_1, it can be written that 1 (Zn^{2+}) × 4 = $\underline{1}$ (S^{2-}) × 4 and the fact that ions have the same coordination is confirmed.

The structure can be built by giving the coordinates of Zn^{2+} ions (0, 0, 0) and S^{2-} ions (¼, ¼, ¼) or the inverse, Zn^{2+} ions (¼, ¼, ¼) and S^{2-} ions (0, 0, 0), and the point group of ZnS is F -43m. The number of formula units in the unit cell (Z) is 4 as there are at the vertices 8 × 1/8 and on the faces 6 × 1/2 or four anions and four cations (Z = 4) and the coordination of ions is 4:4.

ZnS takes the form of two F packings shifted by a quarter of the diagonal of the cube or a cubic F packing of which half of the tetrahedral sites are filled by ions of opposite charge.

3.3.4.4. Compounds of ZnS (sphalerite) type

The compounds of ZnS (sphalerite) type are formed (see Table 3.12) of polarizing cations (Cu^+, Ag^+, etc.) with 1 or 2 charges (Cd^{2+}, Hg^{2+}, etc.) that yield halides and sulfides, selenides and tellurides.

Compounds (MX)	R^+; R^-	a	Ionicity	R^+/R^-	Avail. (pm)
BeSe [WYC 63b]	27; 198	507.0	23%	0.136	−5.5
BeS [WYC 63b]	27; 184	485.0	23%	0.147	−1.0
BeTe [WYC 63b]	27; 221	554.0	7%	0.122	−8.1
CuCl [BEC 25]	60; 184	550.1	33%	0.331	−2.8
CuBr [WYC 63b]	60; 196	569.1	26%	0.306	−9.6
MnS [WYC 63b]	66; 184	560.0	24%	0.359	−7.5
MnSe [WYC 63b]	66; 198	582.0	24%	0.333	−12.0
ZnS [YIM 69]	60; 184	547.5	20%	0.326	−6.9
ZnSe [WYC 63b]	60; 198	566.9	20%	0.303	−12.5
CuI [KRU 52]	60; 220	615.0	15%	0.273	−13.7
ZnTe [HOL 68]	60; 221	610.2	5%	0.271	−16.8
CdTe [WYC 63b]	78; 221	648.0	4%	0.353	−18.4
CuF [WYC 63b]	60; 133	608.5	67%	0.451	$-8.8/_{+2.2}$
CdS [WYC 63b]	78; 184	581.8	19%	0.424	$-10.1/_{+7.9}$
AgI [WYC 63b]	100; 220	647.3	14%	0.455	$-39.7/_{+8.3}$
HgS [WYC 63b]	96; 184	585.2	12%	0.522	$-26.6/_{+17.4}$
HgSe [WYC 63b]	96; 198	581.8	12%	0.485	$-30.5/_{+13.5}$
HgTe [DEL 63]	96; 221	561.1	1%	0.434	$-37.3/_{+10.7}$

Table 3.12. *Several compounds of ZnS (sphalerite) type*

The compounds are listed in three series depending on the value of the steric criterion and it can be noted that it corresponds to coordination 4 only for 50% of them. One part (17%) presents a lower value and the other a higher value (33%).

The hard sphere model applies to all the compounds. Those with the highest ionicity have negative availability that is associated with weak ionicity, which confirms the strongly covalent nature of their bond.

For the others, the part of covalence leads to a negative availability (Avail.), especially when the steric criterion is above the expected value (for large cations Ag^+ and Hg^{2+}).

It should be noted that anionic coordination is negative for the first three compounds (−18.8 pm, −12.6 pm and −25.2 pm, respectively) and goes in the direction of a certain covalence that imposes, in the hard sphere model, the decrease in the size of ions.

Ionicity is the dominant criterion, as it is sufficient to propose a significant part of covalence for these compounds.

The steric criterion is ineffective, as they cannot be expected to pass to coordination 6, which increases isotropy and corresponds to a mainly ionic bond.

Compounds (MM')	R_M; $R_{M'}$	a	Modified availability (pm)
GaP [WYC 63b]	123; 109	545.1	+4.0
BP [WYC 63b]	84; 109	453.8	+3.5
InP [WYC 63b]	142; 109	586.9	+3.1
GaSb [WYC 63b]	123; 140	611.8	+1.9
AlSb [WYC 63b]	124; 140	613.5	+1.6
BN [WYC 63b]	84; 71	361.5	+1.5
SiC [WYC 63b]	114; 75	434.8	−0.7
InSb [MOL 65]	142; 140	647.0	−1.8
BAs [WYC 63b]	84; 125	477.7	−2.1
GaAs [WYC 63b]	123; 125	565.3	−3.2
InAs [WYC 63b]	142; 125	603.6	−5.6
AlAs [WYC 63b]	124; 125	562.0	−5.6

Table 3.13. *Several non-ionic compounds of ZnS (sphalerite) type*

For low ionicity compounds (see Table 3.13), the use of ionic radii is not an option, as the bonds are strongly covalent. Indeed, these are mainly semiconductors, to which the ionic model is not applicable.

The modified availability can be calculated on the basis of covalent radii, and the values of tolerance are consistent with the hard sphere model.

The bond has therefore a strong covalent character and the atoms are packed in such a way that the distance between them is close to that adopted by them in the covalent bond. Although not a proof, this calculation is consistent with a strongly covalent bond.

3.3.5. *Study of zinc sulfide (wurtzite)*

A further structure of MX type that emerges from the series that has been described is that of ZnS (wurtzite), with the same empirical formula as the previous ones, but with a different spatial arrangement (see Figure 3.42). Indeed, the conventional unit cell is no longer cubic, but hexagonal close-packed (1 and 2), the elementary rhombohedral unit cell is represented in (3) but it is not used.

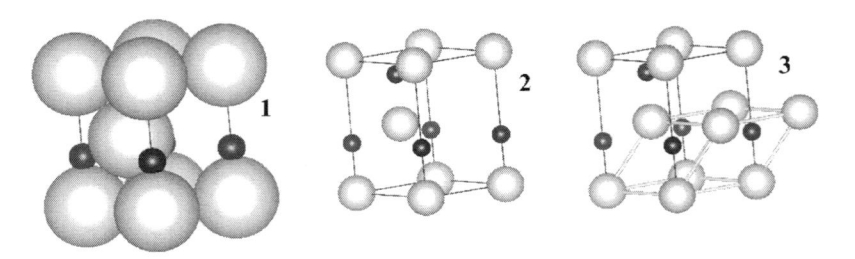

Figure 3.42. *Representation of the conventional and simple SZn (wurtzite) unit cell*

3.3.5.1. *Description of ZnS (wurtzite) based on ion lattices*

As shown in Figure 3.43, SZn crystal (1 and 3) and ZnS crystal (5 and 7) can be described by the imbrication of two hexagonal close-packed structures, one of S^{2-} ions (2 and 4) and the other of Zn^{2+} ions (6 and 8) shifted by 3/8 of c materialized in (7).

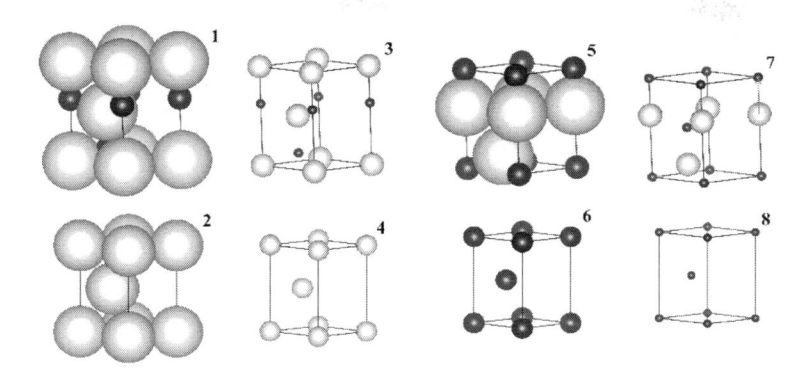

Figure 3.43. *Conventional unit cells of SZn and ZnS (wurtzite)*

The size difference between the two ions leads to the SZn unit cell appearing as larger than that of ZnS, but they are identical (see Figure 3.43).

Cations (see Figure 3.44) form anti-cuboctahedra (1) and so do anions (3), since hexagonal close-packed structures are involved, as mentioned during the description of the unit cell (2). Two anti-cuboctahedra (4) have been represented in order to visualize their relative positioning.

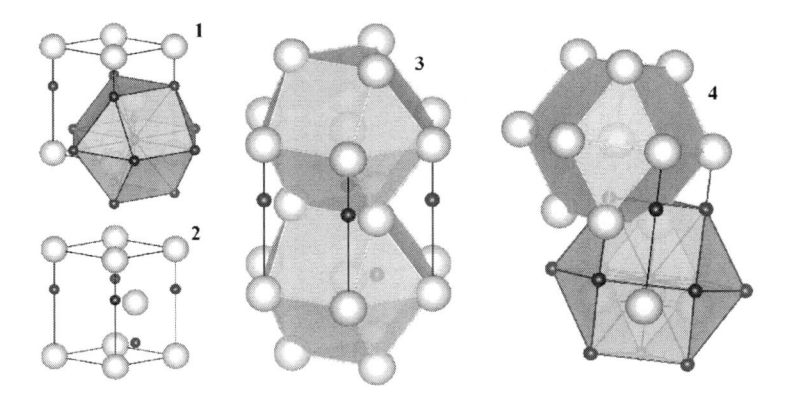

Figure 3.44. *Coordination polyhedra of identical ions in SZn (wurtzite). For a color version of this figure, see www.iste.co.uk/valls/inorganic.zip*

The two anti-cuboctahedra are shifted by c/8 along the axis c and by a/3 and 2a/3 along the other axes, or the inverse, depending on their positioning. Comparisons of ZnS sphalerite (1) and ZnS wurtzite (2) can be drawn in order to observe that sphalerite is formed of two packings of shifted cuboctahedra (see Figure 3.45), therefore of two cubic F packings.

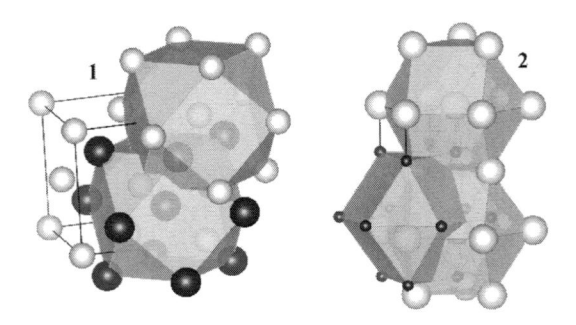

Figure 3.45. *Relative positioning of the coordination polyhedra of identical ions in SZn sphalerite and wurtzite. For a color version of this figure, see www.iste.co.uk/valls/inorganic.zip*

Wurtzite is made of two packings of shifted anti-cuboctahedra and therefore of two hexagonal close-packed structures. The cuboctahedra lead to a cubic unit cell and the anti-cuboctahedra to a hexagonal unit cell.

3.3.5.2. *Description of the ZnS (wurtzite) unit cell based on coordination polyhedra*

The coordination polyhedron formed of cations around anions is a tetrahedron with one S^{2-} ion at the center and the length of each edge is equal to a, the lattice parameter. The same is valid for the cations in a tetrahedron of anions (ZnS_4).

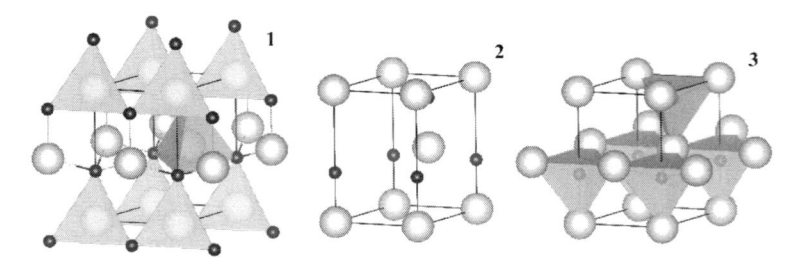

Figure 3.46. *Presentation of the SZn wurtzite unit cell based on ZnS_4 and SZn_4 tetrahedra. For a color version of this figure, see www.iste.co.uk/valls/inorganic.zip*

This tetrahedron can be used to represent the unit cell (2) based on SZn (1), and a correct choice of several ions outside the unit cell allows for a visualization (see Figure 3.46) of the two types of tetrahedra (1 and 3). The SZn₄ tetrahedra (1) have a common ion, which is at the center of a ZnS₄ tetrahedron (3) or its inverse.

The crystal can be described by an aligned packing of tetrahedra that have a common atom (see Figure 3.47), but in order to visualize the symmetries, the unit cell is preferable.

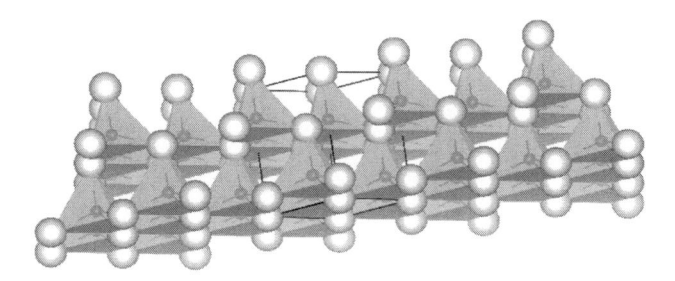

Figure 3.47. *Presentation of SZn wurtzite based on SZn₄ tetrahedra*

Tetrahedra and their alignment are visualized by representing them in large number around the unit cell.

3.3.5.3. *Calculation of various values of ZnS (wurtzite) unit cell*

Ionicity results are identical to those of sphalerite type, the bond being weakly ionic (20%).

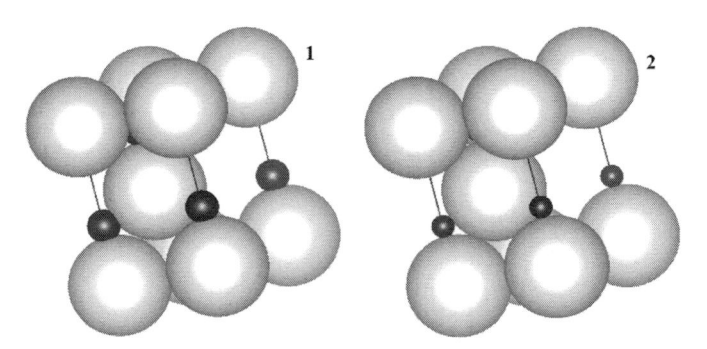

Figure 3.48. *SZn (wurtzite) unit cell and ideal unit cell with all ions in contact*

Ionic radii being identical to those of sphalerite, the calculations yield overall the same results. Similarly, the hard sphere model is appropriate for describing this crystal, if it is considered to be weakly ionic.

Similarly to sphalerite, the cation whose size exceeds the theoretical value (2) has moved away the anions (1) and consequently lowered the packing density (see Figure 3.48).

Counting the fractions of cations in the unit cell shows that there are four anions at four vertices at 60° that are in a proportion of 1/12 in the unit cell and four anions at four vertices at 120° that are in a proportion of 1/6 in the unit cell, and then at the center one anion that is entirely in the unit cell, which yields $4 \times 1/12 + 4 \times 1/6 + 1$ or two cations. A similar counting can be done for the anions, and the number of formula units Z that can be proposed is 2.

Ion coordination can be proposed, which is given in cation:anion order as 4:4 and the point group is P 6_3mc.

ZnS crystal takes the form of two hexagonal close-packed structures shifted by 3/8 of the lattice parameter c or a hexagonal close-packed structure, half of the tetrahedral sites of which are occupied by ions of opposite charge.

3.3.5.4. *Compounds of ZnS (wurtzite) type*

Compounds of ZnS (wurtzite) are formed (see Table 3.14) starting from cations (Ag^+, etc.) with 1 charge or 2 charges (Be^{2+}, Mn^{2+}, Zn^{2+}, etc.) that yield halides and sulfides or selenides and tellurides, respectively.

The compounds are listed by groups, depending on the value of the steric criterion.

The first compound does not follow the steric criterion (too weak), has strong ionicity and weak availability, which allows it to be compatible with the hard sphere model, and confirms the ionic character of the compound.

Compounds (MX)	R⁺; R⁻	a	c	Ionicity	R⁺/R⁻	Avail. (pm)
BeO [WYC 63b]	27; 140	269.8	437.7	61%	0.193	−1.8
MnS [WYC 63b]	66; 184	397.6	643.2	24%	0.359	−6.5
MnSe [WYC 63b]	66; 198	412.0	672.0	24%	0.333	−11.7
ZnSe [WYC 63b]	60; 198	398.0	653.0	20%	0.303	−14.3
ZnS [KIS 89]	60; 184	382.3	625.7	20%	0.326	−9.9
CdSe [WYC 63b]	78; 198	430.0	702.0	19%	0.394	−12.7
ZnTe [WYC 63b]	60; 221	427.0	699.0	5%	0.271	−19.4
ZnO [SCH 14]	60; 140	325.0	520.7	57%	0.429	−1.0
CdS [WYC 63b]	78; 184	413.5	674.9	19%	0.424	−8.8
AgI [HUL 99]	100; 220	459.9	752.4	14%	0.455	−38.4/₊9.7
NH₄F [ZAC 26]	148; 133	439.0	702.0	–	1.053	−12.2
AlN [WYC 63b]	/ 124; 71	311.1	497.8	38%	–	−4.9
InN [WYC 63b]	/ 142; 71	353.3	569.3	31%	–	+3.0
GaN [MIW 93a]	/ 123; 71	314.6	512.5	30%	–	−1.4
SiC [WYC 63b]	/ 114; 75	307.6	504.8	10%	–	−0.5
FeSb [WYC 63b]	/ 140; 124	406.0	513.0	1%	–	−21.0
AgI type NaCl	128; 220	606.7	–	14%	0.582	−44.7
ZnO type NaCl	74; 140	428.0	–	57%	0.529	+0.0

Table 3.14. *Several compounds of ZnS (wurtzite) type*

On the contrary, the five following compounds satisfy the steric criterion of coordination 4, but availability (Avail.) is negative and ionicity is weak.

The hard sphere model is consistent if these compounds are assigned a part of covalent character.

For the next four, the value of the steric criterion is above the expected value, but a change in coordination that would allow them to become compounds of NaCl type is not possible. Only ZnO [BAT 62] and AgI [WEI 62] exist in coordination 6, but at high pressure, which confirms the limitation linked to the energetic criterion.

For the last five, ionic radii cannot be used and the steric criterion makes no sense. Corrected availability is consistent for these compounds too, so a significant part of covalence can be proposed.

> NOTE 3.16.– ZnS sphalerite and wurtzite are neighboring types and they differ by the packing of ions, which is ABC and AB, respectively, and many compounds of these two types are obtained (ZnS, ZnSe, CdS, HgS, HgSe, HgTe).
>
> Several compounds exist under the three types: ZnS sphalerite and wurtzite as well as NaCl (MnSe, ZnTe) and have low ionicity. Some exist as NaCl only under very high pressure (CuCl [HUL 94], CuBr [WYC 63b], MnS [WYC 63b], ZnO [BAT 62], AgI [WEI 62]), which indicates that the energetic criterion is favorable to ZnS type.
>
> The compounds can crystallize in other types, and when pressure is applied, AgI can yield the CsCl type. There are compounds that do not yield the two types of ZnS sphalerite and wurtzite (BeO, CdSe, CdTe, CuF, BeS, BeSe, BeTe, NH4F).
>
> Beryllium compounds have a very weak steric criterion which, on the contrary, is strong for NH4F. CuF has the highest ionicity and CdTe the lowest one, CdSe being the only one with an average value (research may reach the ZnS sphalerite type… or not). All the others have extreme values for one of the criteria and, without offering individual justification, it can be considered that this does not allow them to yield other types.
>
> It can also be reminded that NaCl [SHI 03] and RbCl [WYC 63b] that crystallize as NaCl may yield the CsCl type under very high pressure.
>
> There is very wide diversity but covering it either requires specific conditions of pressure and temperature, or the compound may be unstable, or the field of possibilities has not yet been explored.

In all the compounds, with several exceptions, there is on average no place for the cation at the center of the tetrahedron formed by the anions, and this confirms the strongly covalent character of the bond. The hard sphere model is adapted but only one compound of high ionicity is available for each group (BeO, ZnO and MnS), which validates all the criteria and of BeO that does not validate the steric criterion. Similarly to sphalerite, covalence is the dominant criterion and a change of coordination cannot be taken into account.

3.3.6. *Study of nickel arsenide*

The last structure of MX type described is nickel arsenide (NiAs), which has a hexagonal conventional unit cell (1 and 2) similar to ZnS wurtzite (see Figure 3.49). As shown by the representation of AsNi unit cell (3 and 4), the atoms do not have equivalent environments, which is incompatible with the isotropy of the ionic bond. Therefore, these compounds cannot be considered ionic compounds.

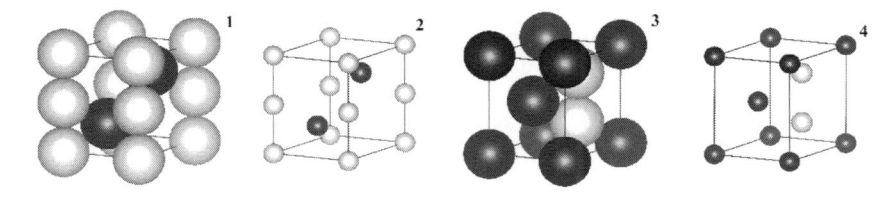

Figure 3.49. *Representation of NiAs and AsNi unit cell*

The modified availability that has been calculated with atomic radii (1) is used, which allows for consistent representation of NiAs unit cell, in contrast with the use of Van der Waals radii (3), and the metallic radii (2) are the ones that lead to perfect contact between nickel atoms (see Figure 3.50).

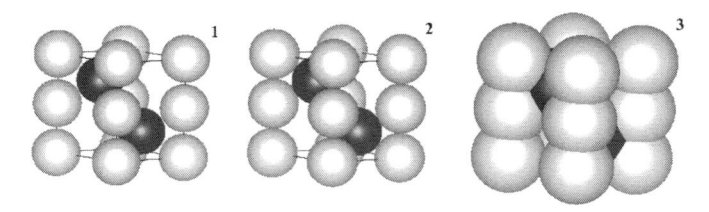

Figure 3.50. *Presentation of NiAs unit cell with covalent radii (1), Van der Waals radii (2) and ionic radii (3)*

3.3.6.1. *Description of the unit cell based on ion lattices*

The conventional NiAs unit cell (see Figure 3.51) is hexagonal (1 and 2), and the nickel atoms form a hexagonal lattice (3), in which arsenic atoms occupy half of the trigonal prism sites (see Figure 3.52). Arsenic atoms form a hexagonal close-packed structure (see Figure 3.51), which is identified by its anti-cuboctahedra (6), in which nickel atoms (6) occupy octahedral sites (see Figure 3.52).

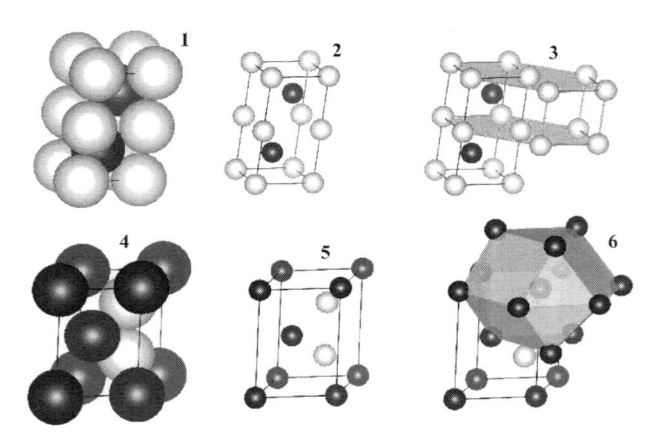

Figure 3.51. *Representation of NiAs and AsNi unit cells. For a color version of this figure, see www.iste.co.uk/valls/inorganic.zip*

The coordination of nickel atoms, at the center of a trigonal prism, allows them to be joined by a metallic bond along the c axis. Moreover, all the tellurides are strongly metallic, as are many compounds of this series.

NOTE 3.17.– A simultaneous representation of the two atom-lattices facilitates the visualization of the imbrication between nickel atoms hexagon and arsenic atoms cuboctahedron.

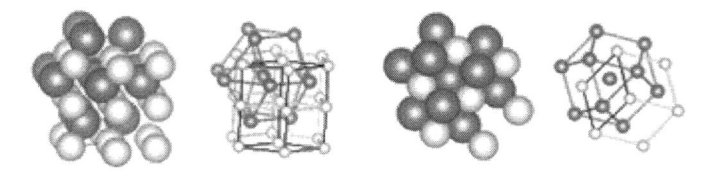

For a visualization of mixed coordination polyhedra, the reader is referred to Figures 3.52 and 3.53.

3.3.6.2. Description of the NiAs unit cell based on coordination polyhedra

The coordination polyhedron in AsNi (1) formed by arsenic atoms around nickel is an octahedron (2) that has horizontally a common edge (3) with the other octahedra (see Figure 3.52) and hence vertically a common face (4).

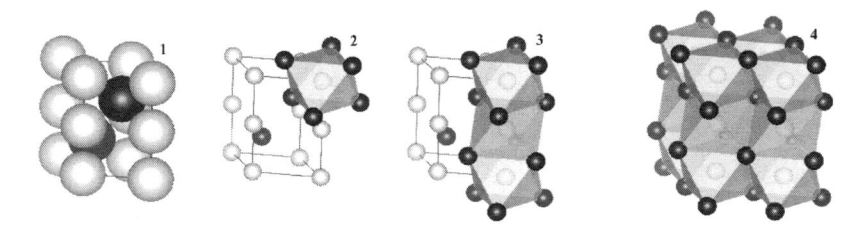

Figure 3.52. Presentation of NiAs unit cell based on NiAs6 octahedra. For a color version of this figure, see www.iste.co.uk/valls/inorganic.zip

The coordination polyhedron (see Figure 3.53) formed by the nickel atoms around arsenic in AsNi (1) is a trigonal prism (2) that has one edge in common with the prisms around it (3).

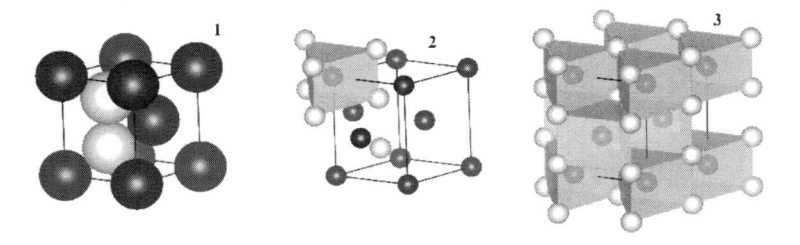

Figure 3.53. Presentation of the AsNi unit cell based on right prisms with triangular base of NiAs^6. For a color version of this figure, see www.iste.co.uk/valls/inorganic.zip

3.3.6.3. Calculation of various values of NiAs unit cell

Electronegativities [HAY 15] of nickel and arsenic are 1.91 and 2.18, respectively; hence, the bond is weakly ionic (2% according to Pauling).

Their covalent radii [HAY 15] are 120 and 117 pm, with lattice parameter a (361.8 pm) and c (503.4 pm) [ALS 25].

188 Inorganic Chemistry

Modified availability can be calculated for the smallest atom (Ni) in the crystal (see Figure 3.54). Considering the triangle ABC in the plane (110) in gray, the length of AB is equal to a quarter of c and AC is at 2/3 of the altitude of triangle BDE (4) or $\frac{2}{3}a\sqrt{3}/2$ or $a/\sqrt{3}$. AB can be calculated according to the Pythagorean theorem as $(a/\sqrt{3})^2 + (c/4)^2$.

Modified availability is +6.9 pm. Although they do not justify the type of bond, covalent radii suggest that a part of covalence is present in the bond.

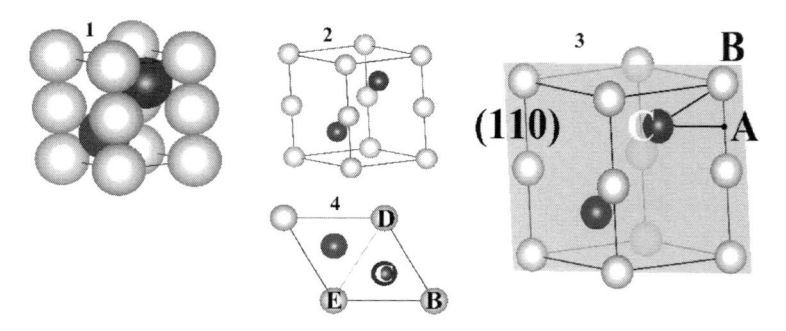

Figure 3.54. *Calculation of the distance between arsenic and nickel atoms in NiAs*

The rule of electrostatic valence is not applicable to NiAs.

Counting the fractions of atoms in the NiAs unit cell leads to two atoms for nickel (4 × 1/12 + 4 × 1/6 + 2 × 2/12 + 2 × 2/6 or 2) and two for arsenic, which are entirely in the unit cell.

AsNi inverse unit cell leads to the same result, that is, 2 for nickel entirely in the unit cell and 2 for arsenic (4 × 1/12 + 4 × 1/6 + 1 or 2), and a number of formula units Z equal to 2 can be proposed.

The structure can be described by considering the space group as P 6_3/mmc and the coordination of atoms is 6:6, but this too reductive way of writing is not sufficient and it is preferable to describe the two coordinations.

3.3.6.4. Compounds of NiAs type

NiAs type of compounds is formed (see Table 3.15) starting from d-block elements that yield sulfides, selenides, antimonides and tellurides.

Compound (MM')	R; R'	a	c	Modified availability (pm)
NiS [KAM 75]	117; 104	343.0	534.0	$+17.8_{/\_}$
FeS [SHE 08]	124; 104	343.0	568.0	$+17.6_{/\_}$
NiSe [KOM 57]	117; 118	365.8	535.4	$+15.0_{/\_}$
FeSe [ALS 25]	124; 118	361.0	587.0	$+12.9_{/\_}$
NiTe [KAM 75]	117; 137	398.0	538.0	$+12.3_{/+4.3}$
PdTe [GRO 56]	130; 137	415.2	567.2	$+11.5_{/+3.5}$
CoS [WYC 63b]	118; 104	336.7	516.0	$+11.3_{/\_}$
PtTe [GRO 55]	130; 137	411.1	544.6	$+6.6_{/-2.4}$
NiAs [ALS 25]	117; 120	361.0	502.8	$+6.4_{/-6.6}$
PtB [STE 60]	130; 84	335.8	408.8	$+4.8_{/\_}$
PtSb [KJE 69]	130; 140	412.6	548.1	$+4.8_{/-10.0}$
NiSb [LE 94]	117; 140	394.6	514.8	$+4.7_{/-9.1}$
CoSb [TER 92]	118; 140	390.0	517.2	$+1.7_{/-11.1}$
VSe [YUR 87]	144; 118	366.0	598.8	$-3.0_{/+10.0}$
TiS [ONO 87]	148; 104	328.0	642.1	$-3.7_{/\_}$
TiSb [MEL 03]	148; 140	410.3	628.4	$-3.7_{/-9.5}$
CrTe [ETO 01]	130; 124	379.0	580.0	$-4.5_{/+0.5}$
VS [BIL39]	144; 104	334.0	578.5	$-7.0_{/\_}$
TiSe [HAH 59]	148; 118	357.5	599.5	$-10.9_{/\_}$
TiTe [MAT 92]	148; 137	386.0	632.5	$-11.7_{/-11.7}$
TaN [FON 69]	158; 71	304.8	491.8	$-14.3_{/\_}$
ZrTe [SOD 79]	164; 137	396.2	669.3	$-17.6_{/-15.6}$
AuSn [WES 31]	130; 140	432.3	552.3	$+15.2_{/+0.2}$
PbPt [ZHU 62]	145; 130	426.0	544.1	$+6.4_{/-20.3}$
PtSn [CHA 70]	130; 140	410.3	542.8	$+3.1_{/-6.4}$
IrSn [MAY 78]	132; 140	398.0	563.0	$-2.5_{/-7.0}$

Table 3.15. *Several compounds of NiAs type*

The compounds are listed in descending order of modified availability, as ionicity has no specific significance since they are mainly covalent or metallic.

The use of metallic radii obtained starting from metals generally in coordination 12 (or 8) is an approximation as, in these compounds, the metal atoms are in hexagonal mode and additionally there is the influence of inserted atoms.

Tendencies can nevertheless be noted, as the compounds proposed in Table 3.15, whose availability is outside the tolerance interval, have availability within tolerance due to the use of metallic radius. An option can be made for a part of metallic character of the bond of these compounds, all the more so as the calculation along the axis c (still with the same tolerance) shows that atoms are in contact.

> NOTE 3.18.– These three typical crystals (CsCl, NaCl and ZnS) have in common a cubic (I or F) packing and a pattern with two atoms oriented respectively along the cube diagonal and equal to the semi-diagonal (1) or oriented along an edge and equal to the half edge (2) or still oriented along the cube diagonal and equal to a quarter of it (3).

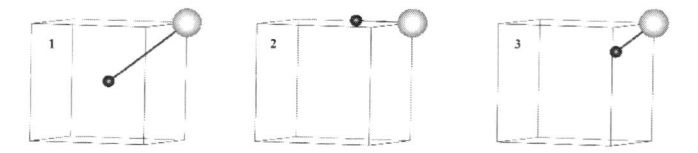

The first and second patterns allow the crystal to have the same symmetry as the lattice but the third one does not allow for preserving the same level of symmetry.

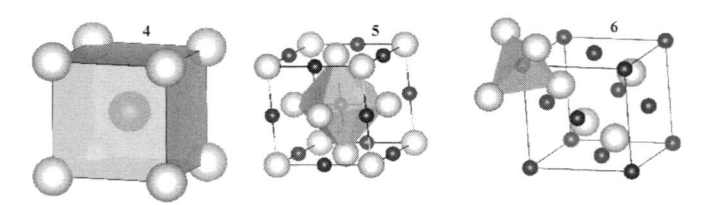

For a color version of this figure, see www.iste.co.uk/valls/inorganic.zip

For a convincing argument, the pattern whose maximal symmetry is compatible with that of the lattice should be defined. The pattern for CsCl is the cube (4), for NaCl it is the octahedron having the same symmetries as the lattice (5), but for ZnS it is the tetrahedron (6) that has no center of symmetry!

A certain number of compounds obey the tolerance regardless of the type of radius and although there are no elements for a conclusion, physical characteristics such as conductivity can be approached.

It is worth remembering that tellurides are generally strongly metallic, and this confirms the previous calculations (very low modified availability with metallic radii) as well as the contact of metal atoms along the c-axis, which allows for conduction.

3.4. Geometry of binary crystals of MX$_2$ type

3.4.1. *Study of calcium fluoride*

3.4.1.1. *Description of the CaF$_2$ unit cell based on ion lattices*

Calcium fluoride CaF$_2$ (1 and 2) and F$_2$Ca (3 and 4) are shown in Figure 3.55, which evidences the packing density of the lattice of fluorine ions. A cubic F packing (1 and 2) of calcium ions of parameter a is visualized, all its tetrahedral sites being occupied by a fluorine ion, as well as a cubic P packing of fluorine ions of parameter a/2, half of the unit cells of which are occupied by a calcium ion.

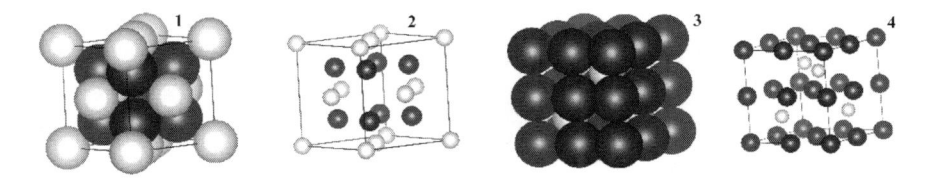

Figure 3.55. *Representation of CaF$_2$ and F$_2$Ca unit cells*

The layer-by-layer construction of the unit cell (1–5) facilitates the visualization of the position of all the atoms (see Figure 3.56).

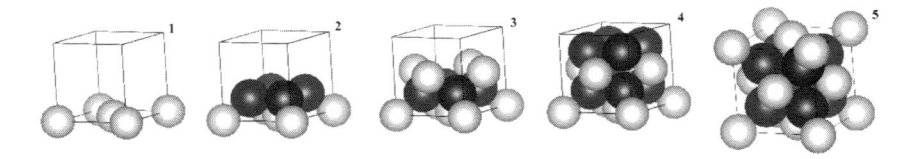

Figure 3.56. *Structuring of CaF_2 unit cell*

In the CaF_2 unit cell (1), the fluorine atoms form octahedra (see Figure 3.57) that have common edges (3) and are characteristic of a cubic P packing (2). The calcium ions form cuboctahedra connected by one face (4) characteristic of a cubic F packing. The two packings have a common center, as the two cubes that represent them are centered.

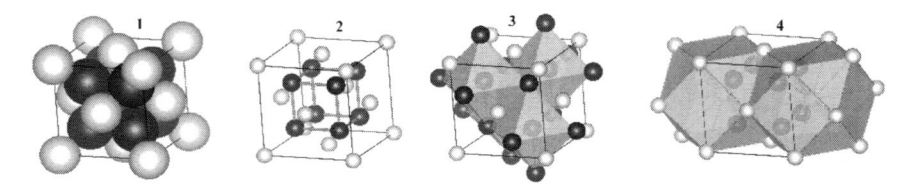

Figure 3.57. *Representation of ion packings in the CaF_2 unit cell. For a color version of this figure, see www.iste.co.uk/valls/inorganic.zip*

3.4.1.2. *Description of the CaF_2 unit cell based on coordination polyhedra*

As it can be seen, in the CaF_2 unit cell (1), each F^- ion is surrounded by four Ca^{2+} ions (see Figure 3.58) and FCa_4 coordination polyhedron is a tetrahedron (1 and 2).

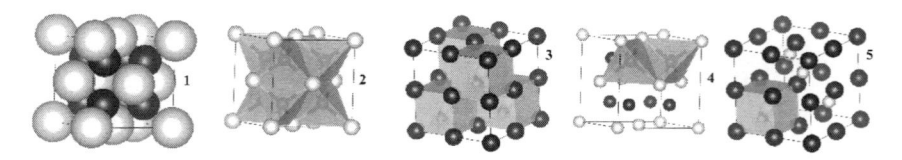

Figure 3.58. *Tetrahedral and cubic coordination polyhedra in CaF_2. For a color version of this figure, see www.iste.co.uk/valls/inorganic.zip*

Similarly, in the F_2Ca unit cell (4), each Ca^{2+} ion is surrounded by eight F^- ions and the CaF_8 coordination polyhedron is a cube (3 and 4).

3.4.1.3. Calculation of various values of the CaF₂ unit cell

Electronegativities [HAY 15] for calcium and fluorine are 1.0 and 4.0, respectively, and the bond is strongly ionic (ionicity 89%).

Ionic radii [HAY 15] for calcium and fluorine are $112^{(VIII)}$ and $133^{(IV)}$ pm, respectively, with a lattice parameter a of 546.2 pm [OFT 27]. The availability for the cation in the F_2Ca crystal (see Figure 3.59) can be calculated by considering one (3) of four CaF_8 cubes (2).

The side of this cube is equal to a/2, and its diagonal is equal to $a\sqrt{3}/2$. It can be noted (3) that the diagonal of this cube is equal to $2R^+ + 2R^-$ or 245 pm, which is slightly above the result of the calculation based on the lattice parameter, which was 236.5 pm. The availability is negative (−8.5 pm) and since it is close to tolerance, the bond can be considered ionic and the hard sphere model is validated.

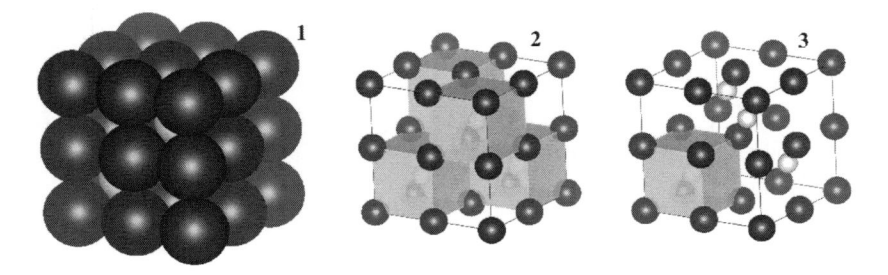

Figure 3.59. *F_2Ca unit cell with cubic coordination polyhedra. For a color version of this figure, see www.iste.co.uk/valls/inorganic.zip*

The calculation of packing density leads to 0.628 with indicated values of a and ionic radii but has no specific meaning other than showing the presence of empty sites in F_2Ca or the spreading of calcium ions required for the inclusion of fluorine ions (see Figure 3.59).

The calculation of distances d_{Ca-F} = 236.5 pm < $R^+ + R^-$ (of 8.5 pm) confirms the ionic bond. Spaces e_{Ca-Ca} = 162.2 pm and e_{F-F} = 7.1 pm show the absence of contact between ions of the same nature.

Applying the electrostatic valence rule, we can verify that if cations have coordination 8, for Ca_1F_2, it can be written that $1(Ca^{2+}) \times 8 = \underline{2}\,(F^-) \times 4$ and this confirms that cations have coordination 4.

The structure can be described by giving the coordinates of the atoms of Ca^{2+} (0,0,0) and (1/2,1/2,0) and those of F^- (1/2,1/2,1/2) or for the inverse unit cell F^- (0,0,0), (1/2,1/2,0) and (1/2,0,0) and Ca^{2+} (1/4,1/4,1/4). Finally, the point group is F m-3m.

A more complete description of CaF_2 results from giving the number of formula units in the unit cell (Z), which is four as there are (8 × 1/8 + 6 × 1/2) Ca^{2+} and 8 F^- and ion coordination is 8:4.

CaF_2 compound takes the form of two centered cubic packings of parameter a and type F for the calcium ions and of parameter a/2 and type P for fluorine ions.

3.4.1.4. Compounds of CaF₂ type

The compounds of CaF_2 type are formed (see Tables 3.16–3.19) of elements that yield ions with 2 or 4 charges (alkaline-earth, d- and f-blocks), which are oxides and halides, respectively.

This first table lists compounds whose steric criterion is below that corresponding to coordination 8. It can be noted that availability is within tolerance, which validates the hard sphere model and allows the proposition of an ionic bond for these compounds (see Table 3.16).

Compounds (MX₂)	R⁺; R⁻	a	R⁺/R⁻	Ionicity	Availability (pm)
TiF₂ [MOR 87]	91; 133	515.6	0.647	78%	−0.8
CeO₂ [WYC 63a]	97; 140	541.1	0.693	76%	−2.7
PrO₂ [WYC 63a]	96; 140	546.9	0.686	75%	+0.8
PuO₂ [WYC 63a]	91; 140	539.6	0.643	71%	+2.7
SrCl₂ [MAR 25]	126; 181	697.7	0.696	71%	−4.9
CmO₂ [WYC 63b]	95; 140	537.2	0.679	70%	−2.4
ZrO₂ [WYC 63a]	84; 140	507.0	0.600	69%	−4.5
NpO₂ [WYC 63a]	92; 140	543.4	0.700	68%	+3.3
UO₂ [WYC 63a]	100; 140	546.8	0.714	56%	−3.2
PbO₂ [LIU 80]	94; 140	520.0	0.671	51%	−8.8

Table 3.16. *Several compounds of CaF₂ type (R⁺/R⁻ < 0.732)*

The formula stoichiometry and the ionic nature of the bond do not allow the passing into coordination 6. There is no combination of ions that can simultaneously satisfy the constraints of the steric criterion, the formula stoichiometry, and maintain a mainly ionic bond.

> NOTE 3.19.– Two compounds exist as other types: PbO_2 whose ionicity is the weakest and shifts into coordination 6, yields a crystal of TiO_2 type [BOL 97] and uranium oxide that shifts into coordination 4 of ZnS sphalerite type [LAU 89] but with strong evolution of stoichiometry ($UO_{2.338}$).

The second table lists compounds whose steric criterion corresponds to coordination 8 (see Table 3.17) and it can be noted that availability is still within tolerance.

Several fluorides are out of the tolerance interval but the value of ionic radius in coordination 6 has been used for fluorine, whose structure is 8-coordinate, which may explain this slight deviation (of ~2%).

The hard sphere model, as well as Pauling's first rule, are applicable, all the more so as no compound crystallizes under another crystalline type.

Compounds (MX_2)	R^+; R^-	a	R^+/R^-	Ionicity	Availability (pm)
BaF_2 [WYC 63a]	142; 133	620.0	1.068	91%	−6.5
SrF_2 [WYC 63a]	126; 133	580.0	0.947	90%	−7.9
CaF_2 [OFT 27]	112; 133	546.2	0.842	89%	−8.5
EuF_2 [WYC 63a]	125; 133	579.6	0.940	86%	−7.0
CdF_2 [WYC 63a]	110; 133	538.8	0.827	74%	−9.7
$BaCl_2$ [WYC 63a]	142; 181	734.0	0.785	72%	−5.2
ThO_2 [WYC 63a]	105; 140	560.0	0.750	70%	−2.5
PbF_2 [ACH 05]	129; 133	594.0	0.970	70%	−4.8
HgF_2 [EBE 33]	114; 133	554.0	0.857	67%	−7.1

Table 3.17. *Several compounds of CaF_2 type ($0.732 < R^+/R^- < 1$)*

Several intermetallic compounds are proposed (see Table 3.18) and modified availability is used (either with the covalent radius or with the metallic radius).

The intermediate character of this bond is confirmed, partly covalent and partly metallic, supported by the shift of sign upon the change of type of radius.

However, this is just an observation, as this case is beyond the models.

Compounds (MM'$_2$)	R; R'	a	Modified availability (pm)
AuSb$_2$ [WYC 63a]	130; 140	665.6	+18.2/$_{-24.8}$
Mg$_2$Ge [WYC 63a]	120; 140	637.8	+16.2/$_{-7.8}$
AuAl$_2$ [WYC 63a]	130; 124	600.0	+5.8/$_{-28.2}$
PbMg$_2$ [WYC 63a]	145; 140	683.6	+11.0/$_{-42.0}$
AuIn$_2$ [WYC 63a]	130; 142	650.2	+9.5/$_{-3.5}$
AuGa$_2$ [WYC 63a]	130; 123	606.3	+9.5/$_{-4.5}$
PtSn$_2$ [WYC 63a]	130; 140	642.5	+8.2/$_{-1.8}$
PtAl$_2$ [WYC 63a]	130; 124	591.0	+1.9/$_{-26.1}$

Table 3.18. *Several intermetallic compounds of CaF$_2$ type*

For the compounds of fluorite type, it can be noted that formula stoichiometry and ionicity are predominant but the hard sphere model is perfectly adapted (although some of them disregard the steric criterion).

3.4.2. Study of lithium oxide

3.4.2.1. *Description of Li$_2$O unit cell based on ion lattices*

Lithium oxide Li$_2$O (3 and 4) is represented in Figure 3.60, together with OLi$_2$ (1 and 2). Characteristics similar to those of CaF$_2$ unit cell are visualized but the ions have reversed their positions and charges. The cation has indeed 2 charges and is located in a cubic site and the anion has one charge in a tetrahedral site. This crystal is qualified as anti-fluorite and belongs to F m-3m space group.

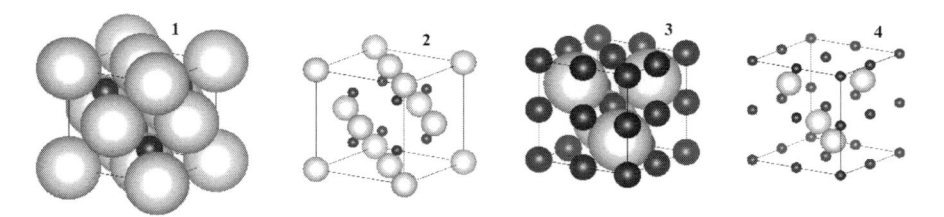

Figure 3.60. *Representation of OLi$_2$ and Li$_2$O unit cells. For a color version of this figure, see www.iste.co.uk/valls/inorganic.zip*

The steric criterion suggests coordination 4 for lithium ion (0.421) and LiO$_4$ coordination polyhedron is a tetrahedron (1).

3.4.2.2. *Calculation of various values of Li$_2$O unit cell*

Electronegativities for calcium and oxygen are 0.98 and 3.5, respectively [HAY 15]; therefore, the bond is strongly ionic (ionicity 80%).

The ionic radii [HAY 15] for oxygen$^{(VIII)}$ and fluorine$^{(IV)}$ are 142 and 59 pm, respectively, with a lattice parameter a of 468.9 pm [OFT 27]. The availability of the cation in the crystal can be calculated (see Figure 3.60) by considering a cube in Li$_2$O (3). A quarter of the diagonal is equal to the anion–cation distance ($\frac{a}{4}\sqrt{3}$) or 203.0 pm. It can be noted (3) that this value is close to R$^+$ + R$^-$ or 201.0 pm and the availability of +2.0 pm is weak; therefore, the bond is ionic and the hard sphere model is validated.

The spaces e_{Li-Li} = 116.5 pm (a/2) and e_{O-O} = 47.6 pm (a/$\sqrt{2}$ – 2R$^-$) show the absence of contact between ions of the same nature.

The application of the electrostatic valence rule makes it possible to verify that if the coordination of cations is 4, Pauling proposes for the compound Li$_2$O$_{\underline{1}}$: 2(Li$^+$) × 4 = $\underline{1}$(O^{2-}) × 8, which confirms that cations have coordination 8.

The description of Li_2O can be completed by providing the number of formula units in the unit cell (Z), which is 4, since there are (8 × 1/8 + 6 × 1/2) O^{2-} and 8 Li^+. Finally, the coordination of ions is 4:8.

Li_2O takes the form of two centered cubic packings of lattice parameter a and type F for oxygen and of parameter a/2 and of type P for lithium ions.

3.4.2.3. *Compounds of Li_2O type*

The compounds of Li_2O type are formed (see Tables 3.19 and 3.20) on the basis of elements that yield ions with 2 or 4 charges (alkaline-earth, d- and f-blocks) and that are oxides and halides, respectively.

As previously, for several examples of non-ionic compounds, a shift of the sign of modified availability can be noted when metallic radii are used. This is in fact a bond that is partly metallic and partly covalent.

Compounds (M_2M')	M; M'	a	Modified availability (pm)
Mg_2Si [WYC 63a]	114; 140	639.1	+22.7/+1.3
Mg_2Ge [WYC 63a]	120; 140	637.8	+16.2/−7.8
Mg_2Sn [WYC 63a]	140; 140	676.5	+12.9/−9.1
Mg_2Pb [WYC 63a]	145; 140	683.6	+11.0/−20.0

Table 3.19. *Several non-ionic compounds of Li_2O type*

Only several compounds, among which is that of Li_2O type, present a steric criterion that corresponds to coordination 4, the others present a steric criterion corresponding to coordination 6 or 8. It can also be noted that if cation coordination is 6, that of anion should be 12 and, if it is 8, that of cation should be 16 and, in both cases, there are no typical crystals with these characteristics. The steric criterion can no longer play a role and therefore this is the only type observed for these compounds that belong all to the F m-3m space group.

For all these compounds, availability is within tolerance, which confirms their mainly ionic character.

Compounds (M₂X)	R⁺; R⁻	a	R⁺/R⁻	Ionicity	Availability (pm)
Li₂O [FAR 91]	59; 142	462.0	0.415	80%	+2
Li₂Se [WYC 63a]	59; 198	600.5	0.298	48%	+3.0
Li₂S [WYC 63a]	59; 184	570.8	0.321	48%	+4.2
Li₂Te [ZIN 34]	59; 221	650.4	0.267	27%	+1.6
K₂O [ZIN 34]	137; 142	643.6	0.965	83%	−0.3
Rb₂O [WYC 63a]	*150*; 142	674.0	1.056	83%	−0.1
Na₂O [WYC 63a]	99; 142	555.0	0.697	81%	−0.7
K₂Se [ZIN 34]	137; 198	767.6	0.692	55%	−2.6
K₂S [ZIN 34]	137; 184	739.1	0.745	55%	−1.0
Rb₂S [WYC 63a]	*150*; 184	765.0	0.815	55%	−2.7
Na₂Se [KHI 77]	99; 198	680.9	0.500	50%	−2.2
Na₂S [BON 92]	99; 184	654.7	0.538	50%	+0.5
K₂Te [SEI 02]	137; 221	823.3	0.620	34%	−1.5
Na₂Te [WYC 63a]	99; 221	731.4	0.448	29%	−3.3

Table 3.20. *Several compounds of Li₂O type*

NOTE 3.20.– Coordination is 8 for Ca^{2+} cations of fluorite (0.842), but 4 for Li^+ cations of anti-fluorite (0.421). Negative ions surround small positive ions in sites that are cubic and tetrahedral, respectively. There is inversion of both the position of ions and the stoichiometry of the formula.

3.4.3. *Study of rutile*

3.4.3.1. *Description of TiO₂ unit cell based on ion lattices*

It is the first unit cell described that contains an irregular polyhedron. The TiO6 octahedron (see Figure 3.61) has an axis along the (110) plane, represented in blue (3), which is the longest, and the anions present on it are qualified as axial. The four other anions of the octahedron are qualified as equatorial. It can be noted that equatorial atoms do not form a square, but a rectangle whose larger side is equal to the lattice parameter c of the unit cell or 295.8 pm and whose smaller side is equal to 253.0 pm. The distance between axial equatorial anions is 277.9 pm.

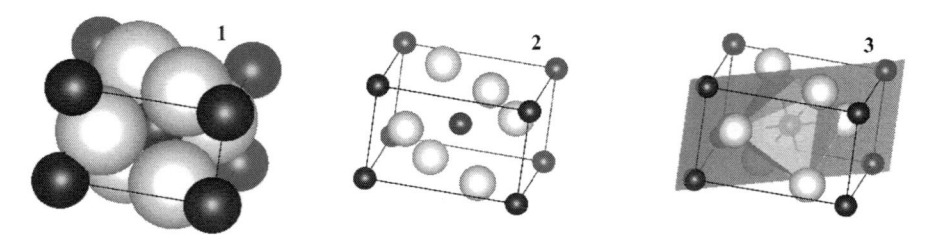

Figure 3.61. *Representation of TiO_2 unit cell. For a color version of this figure, see www.iste.co.uk/valls/inorganic.zip*

In rutile (1), the titanium ions form a tetragonal I lattice (2) and the oxygen ions form a distorted hexagonal close-packed structure (3), whose anti-cuboctahedron (also distorted) is shown in Figure 3.62.

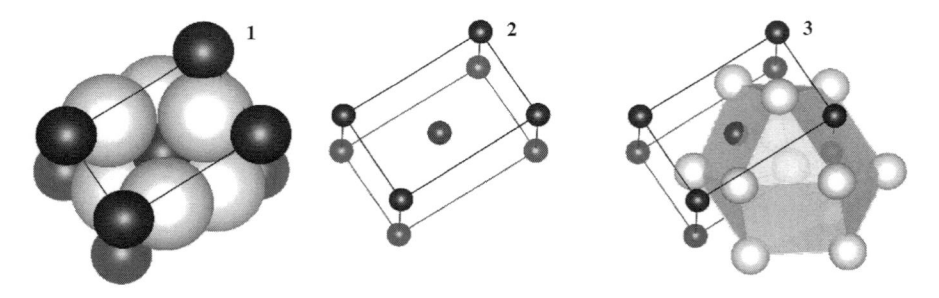

Figure 3.62. *Representation of coordination polyhedra of identical ions in the TiO_2 unit cell. For a color version of this figure, see www.iste.co.uk/valls/inorganic.zip*

The steric criterion corresponds to coordination 6 but rutile must also follow formula stoichiometry, a coordination 3 for the anion and an ionic covalent bond between titanium and oxygen. All these conditions are best satisfied by this structure.

3.4.3.2. Description of TiO₂ unit cell based on coordination polyhedra

Rutile (2) is quadratic (a = 459.4 pm, c = 295.8 pm [WYC 63a]), belongs to space group P 4_2/mnm and the unit cell presents a distorted TiO_6 octahedron (1) inside a centered tile (3) of Ti^{4+} ions (see Figure 3.63). The lattice parameters and the space group do not allow for defining the position of atoms; a parameter x should be set, which corresponds to coordinates x and y (z = 0) of the oxygen atom (the other oxygen atoms are positioned by symmetries). The generally accepted value is 0.305 or 0.31 and it is used for all the compounds.

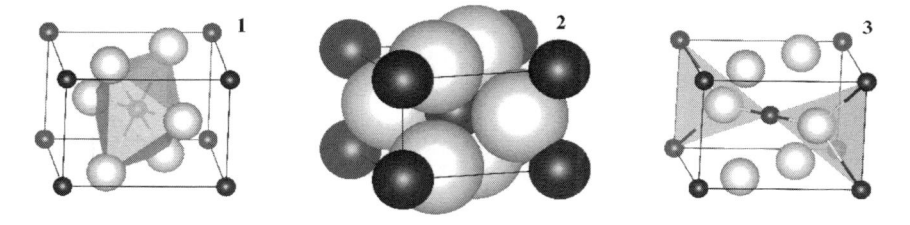

Figure 3.63. *Representation of coordination polyhedra in TiO₂ unit cell. For a color version of this figure, see www.iste.co.uk/valls/inorganic.zip*

3.4.3.3. Calculation of various values of the TiO₂ unit cell

Electronegativities for titanium and oxygen are 1.54 and 3.5, respectively [HAY 15]; therefore, the bond is ionic covalent (ionicity 62%).

Ionic radii [HAY 15] for oxygen and titanium are $136^{(III)}$ and $61^{(IV)}$ pm, respectively, with lattice parameters a of 459.4 pm and c of 295.8 pm [WYC 63a]. The lattice parameters facilitate the calculation of availability, but two anion–cation distances (AB and BC) can be noted, and it is the smaller one (BC) that will be taken into account. It can be noted that due to crystal symmetry, AB is equal to CD and BC is equal to AF, which facilitates the determination of corresponding equations (see Figure 3.64).

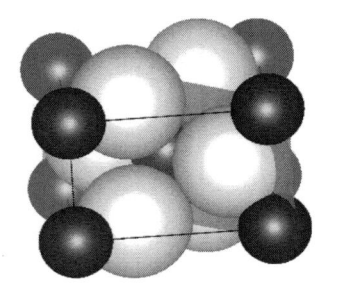

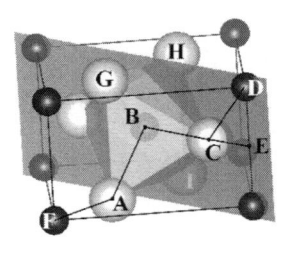

Figure 3.64. *Study of TiO_2 unit cell. For a color version of this figure, see www.iste.co.uk/valls/inorganic.zip*

Availability for the cation in the crystal (see Figure 3.64) can be calculated by considering the shortest distance along AB (B being the central atom), availability of −2.4 pm is within tolerance; therefore, the bond is ionic and the hard sphere model is validated (along BC it is +1.4 pm).

The application of the electrostatic valence rule makes it possible to verify that if the coordination of cations is 6, for Ti_1O_2, $1(Ti^{4+}) \times 6 = \underline{2}(O^{2-}) \times 3$ and this confirms that cations have coordination 3 (see Figure 3.63).

It can be noted that B, C and D are in the (110) plane, which contains the longest axis of the octahedron and AF and BC are equal, and so are AB and CD.

In a regular octahedron, distances between opposite ions are identical, while for rutile two values are observed.

The length of CD (or AB) is obtained by applying the Pythagorean theorem to the CDE triangle:

$$\sqrt{\left((1-2x)\, a\sqrt{2}\right)^2 + \left(\frac{c}{2}\right)^2} \qquad [3.1]$$

The length of AF (or BC) is given by the expression

$$x\, a\sqrt{2} \qquad [3.2]$$

If parameter x is set at 0.305 (value that minimizes availability and spaces between anions), it can be verified that the space between anions is +23.8

along GA (1) and it is fixed by the lattice parameter c, which is +4.2 along GC (1 and 2) and −9.5 pm along GH (2).

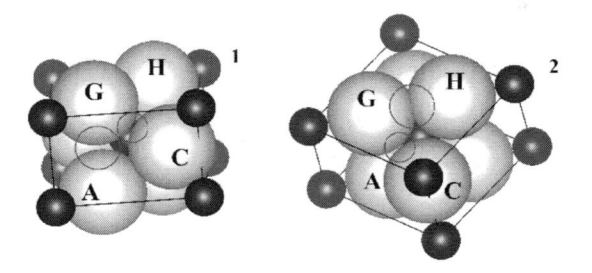

Figure 3.65. *Availability of ions in TiO_2 unit cell. For a color version of this figure, see www.iste.co.uk/valls/inorganic.zip*

The sides of the octahedron vary depending on directions and three values are observed: 295.8 pm (it is parameter c), 277.9 pm and 253.0 pm. This justifies the name of the distorted octahedron.

> NOTE 3.21.– The spaces between anion and cation are −2.4 pm (along AB or CD) or +1.4 pm (along BC or AF), and the spaces between anions are −9.5 pm (along GH or AI), +5.9 (along CA, CG, etc.) or +23.8 pm (along AG or HI). The adjustment of parameter x leads to minimizing certain constraints, but to the detriment of others. The increase in x leads to an increase in the distance between anions along GH, but availability along BC decreases. Rutile has been represented for x = 0.32 (1) and x = 0.28 (2) as well as the best compromise (3) for which tolerances are best respected for all the ions having x equal to 0.305.

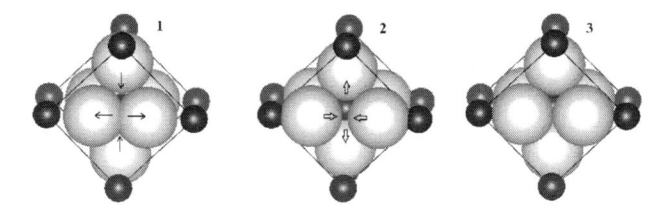

The optimal value of x may slightly vary from one compound to another, and it can be noted that the solid presents certain adaptability in order to meet the constraints that are imposed on it.

The description of TiO$_2$ can be completed by giving the number of formula units in the unit cell (Z), which is 2, since there are (4 × 1/2 + 2) O^{2-} and (8 × 1/8 + 1) Li$^+$ and finally, coordination of ions, which is 6:3 (see Figure 3.64). The steric parameter is 0.449 and corresponds to the expected value of coordination 6.

Rutile (TiO$_2$) takes the form of a centered quadratic packing of titanium ions of parameters a and c and of a distorted hexagonal close-packed structure of oxygen ions. The packing of oxygen ions can also be represented by a distorted octahedron positioned along a diagonal of the centered tetragonal prism formed of titanium ions.

3.4.3.4. Compounds of TiO$_2$ type

The crystals of rutile (TiO$_2$) type relate to oxides of elements of d-block that are susceptible to yield ions with 4 charges (Cr, Mn, Mo, Os, Pb, Sn, W, etc.) and the fluorides yield ions with 2 charges (Fe, Co, Mg, Mn, Ni, Zn, etc.).

Compounds (MX$_2$)	R$^+$; R$^-$	a	c	R$^+$/R$^-$	Ionicity	Avail. c	Avail. a
MnO$_2$ [BOL 93]	53; 136	440.4	287.7	0.390	61%	−0.9	−14.7
CrO$_2$ [BAU 71]	55; 136	442.1	291.7	0.404	57%	−1.0	−14.3
SiO$_2$ [KRO 03]	40; 136	422.4	269.0	0.294	47%	+1.8	−19.7
GeO$_2$ [LOD 01]	53; 136	438.4	438.4	0.390	43%	−1.7	−15.3
IrTe$_2$ [JOB 01]	63; 221	614.7	614.7	0.285	0%	−14.2	−51.7

Table 3.21. *Several compounds of TiO$_2$ type (R$^+$/R$^-$ < 0.414)*

Table 3.21 lists several binary compounds whose steric criterion is below the value corresponding to coordination 6. The cationic availability (Avail. c) along AB and the anionic availability (Avail. a) along GH have been provided in picometers.

The change of type in order to meet the steric criterion is not possible, as CdI$_2$ and CdCl$_2$ types correspond to compounds of weak ionicity.

The hard sphere model is validated, IrTe$_2$ that is strongly covalent is the only one outside the tolerance interval, either along BC direction or along AB direction (availability −18.5 or −14.2 pm) or for anionic availability

(−51.7 pm). However, it should be noted that it only exists under strong pressure (negative availability observed in this case has already been indicated) and takes the form of CdI_2, which validates the hard sphere model.

This is the first series of compounds for which availability between anions is negative, with all the previous types being generally positive.

Compounds (MX_2)	R^+; R^-	a	c	R^+/R^-	Ionicity	Avail. c	Avail. a
MgF_2 [HAI 01]	72; 133	462.1	301.6	0.541	84%	−7.7	−5.8
VF_2 [COS 83]	79; 133	480.3	323.5	0.594	75%	−3.1	−0.8
ZnF_2 [OTO 01]	74; 133	470.3	313.4	0.556	75%	−3.7	−3.5
FeF_2 [BAU 71]	61; 133	469.5	331.0	0.459	69%	+16.0	−3.7
CoF_2 [BAU 71]	65; 133	469.5	317.7	0.489	67%	+6.8	−3.7
NiF_2 [COS 93]	69; 133	465.0	308.4	0.519	66%	−1.6	−5.0
TiO_2 [WYC 63a]	61; 136	459.4	295.8	0.449	62%	−2.4	−9.5
VO_2 [HOE 71]	58; 136	451.7	287.2	0.426	58%	−4.0	−11.6
WO_2 [WYC 31]	66; 136	486.0	277.0	0.485	56%	−9.4	−2.2
PdF_2 [WYC 63a]	86; 133	493.1	336.7	0.647	56%	−2.7	+2.8
PbO_2 [BOL 97]	78; 136	495.8	338.8	0.574	51%	+3.6	+0.5
SnO_2 [BAU 71]	69; 133	473.8	318.7	0.507	45%	+0.9	−5.5
TeO_2 [GOL 26]	97; 136	479.9	377.8	0.713	39%	−2.5	−3.9
MoO_2 [SEI 04]	65; 136	484.7	281.4	0.478	36%	−7.1	−2.5
IrO_2 [BOL 97]	63; 136	450.5	315.9	0.463	34%	+1.8	−12.0
RuO_2 [BOL 97]	62; 136	449.7	310.5	0.456	34%	+0.6	−12.2
OsO_2 [BAU 71]	63; 136	450.0	318.4	0.463	34%	+2.7	−12.1

Table 3.22. *Several compounds of TiO_2 type $(0.414 < R^+/R^- < 0.732)$*

Table 3.22 lists binary compounds whose steric criterion corresponds to the value of coordination 6. It can be noted that availability is within tolerance interval or close. This negative value confirms a certain covalent character in Ti–O bond.

The hard sphere model is applicable and the distances between anions along GH or GC (see Figure 3.65) should be taken into account, in order to analyze all the interactions in the crystal. Indeed, for FeF_2 and CoF_2, for example, since the cationic and anionic availabilities evolve in the same direction, the anion can impose a positive availability to the cation.

3.4.4. Study of cadmium iodide

3.4.4.1. Description of CdI_2 unit cell based on ion lattices

The compounds of cadmium iodide type (CdI_2) are the first ones to have layers of atoms separated by spaces without atoms.

The unit cell is hexagonal (1 and 2) and by indicating an atom outside the unit cell (see Figure 3.66), the octahedron around the cation (3) can be visualized.

If several octahedra of neighboring unit cells are represented, the anion packing can be imagined in layers with one layer out of two without cations (4).

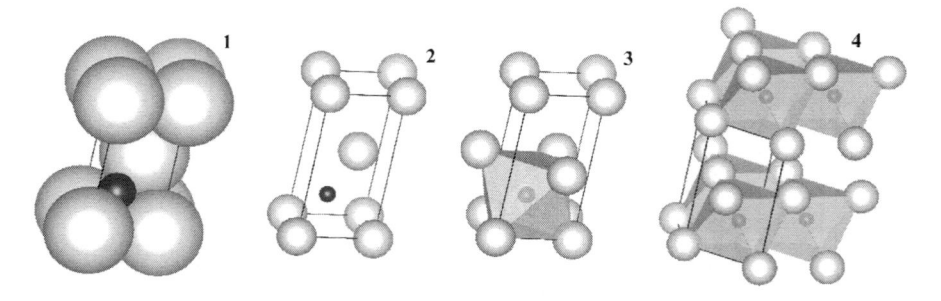

Figure 3.66. *Representation of CdI_2 unit cell*

Certain representations may lead to confusions (see Figure 3.67) as there is no space between layers, but only empty sites on the same level.

The representations in real and unfolded model are therefore essential in order to understand the crystal structure.

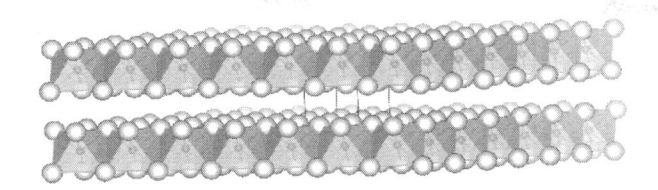

Figure 3.67. Representation of the packing of octahedra in CdI_2

Figure 3.68 shows the planes containing the iodine anions (1) and then cadmium cations.

The interval between two layers of iodine ions is c/2, whereas the interval between the layers of cadmium ions is c (see Figure 3.68).

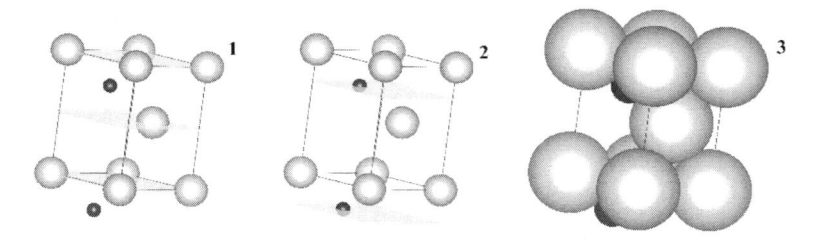

Figure 3.68. Representation of CdI_2 unit cell. For a color version of this figure, see www.iste.co.uk/valls/inorganic.zip

The cations form a hexagonal lattice (1), and the anions a hexagonal close-packed structure (2), which are represented superimposed (3) for a better visualization of their respective positioning (see Figure 3.69).

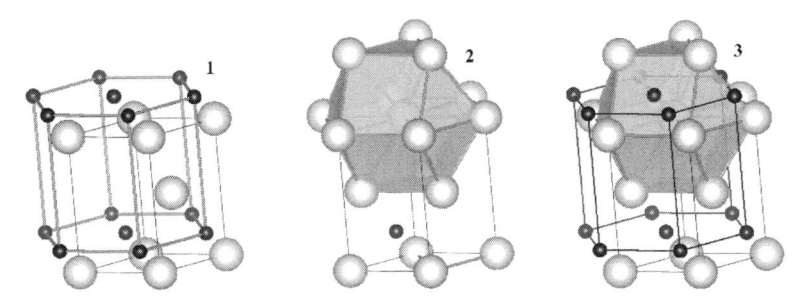

Figure 3.69. Representation of coordination polyhedra in CdI_2. For a color version of this figure, see www.iste.co.uk/valls/inorganic.zip

3.4.4.2. *Description of CdI₂ unit cell based on coordination polyhedra*

The cadmium ion (see Figure 3.70) is in octahedral coordination (1), and the iodine ion is surrounded by three cadmium ions (2).

This arrangement satisfies the electrostatic valence rule and shows the anisotropy of the environment of iodine ions and therefore the more strongly covalent character of bonds.

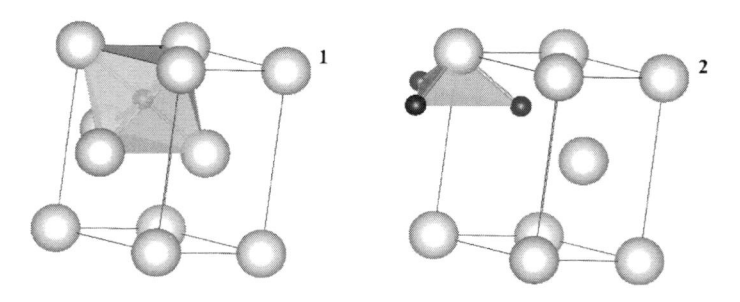

Figure 3.70. *Representation of coordination polyhedra of the CdI₂ unit cell. For a color version of this figure, see www.iste.co.uk/valls/inorganic.zip*

3.4.4.3. *Calculation of various values of CdI₂ unit cell*

Availability results from calculating the distance between the two closest possible opposite ions. Similarly to rutile, in this crystal, coordination polyhedra are not regular and the two lattice parameters have to be taken into account in order to calculate availability.

On the basis of the ABC triangle represented in Figure 3.71, it can be noted that AC is equal to the coordinate along c of this atom or $\frac{c}{4}$.

AB can readily be calculated on the basis of a top view of the unit cell (2), as it is equal to 2/3 of the altitude of the equilateral triangle, which is $\frac{2}{3} a \frac{\sqrt{3}}{2}$ or $\frac{a}{\sqrt{3}}$.

The application of the Pythagorean theorem yields the distance between anion and cation from which the radius of the anion is deducted, which yields the spatial availability

$$\text{availability} = \sqrt{\left(\frac{c}{4}\right)^2 + \left(\frac{a}{\sqrt{3}}\right)^2} - R^- \qquad [3.3]$$

This calculation yields the set of availabilities proposed in Table 3.23.

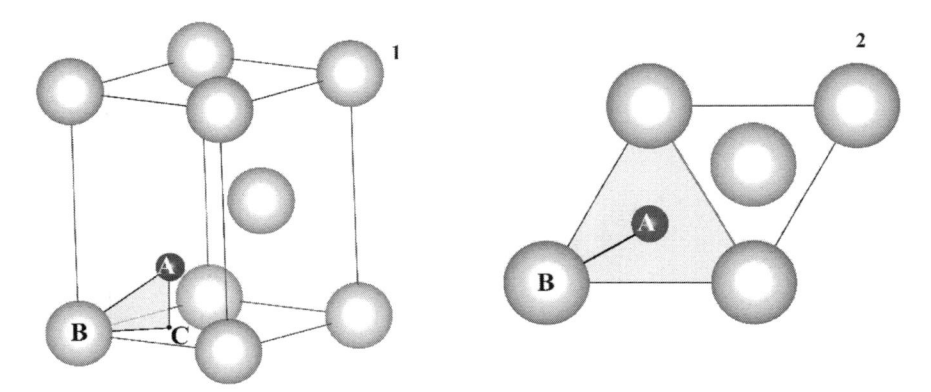

Figure 3.71. *Study of the CdI$_2$ unit cell*

The hard sphere model is not applicable to CdI$_2$, and for example, the lattice parameter a, which should be twice the radius of anions as a condition for the latter to be perfectly tangent, is generally lower than this value. This should be taken into account when interpreting the results.

3.4.4.4. Compounds of CdI$_2$ type

The compounds of CdI$_2$ type are halides, hydroxides, tellurides or selenides of divalent ions and several sulfides of tetravalent ions.

Table 3.23 lists the compounds whose steric criterion is below the expected value for coordination 6. The shift to coordination 4 cannot be envisaged, as no typical compound follows the formula stoichiometry and this coordination.

Ionicity is weak for all these compounds, and they can be assigned a strong covalent character simply on the basis of this criterion.

It is the first series of compounds for which anionic availability is generally negative (the lattice parameter a is below the double anion radius) when the steric criterion is below the expected value. It is consistent only if the bond is assigned a strong covalent character.

The hard sphere model is correct, as availability (Avail.) leads to consistent values, with the exception of several compounds (Fe(OH)$_2$ and FeCl$_2$). The compounds have therefore a strong covalent character that can lead to negative values of availability. Nevertheless, there is a part of ionic bond, as modified availability (calculated with covalent radii) yields inconsistent results.

Compounds (MX$_2$)	a	c	R$^+$/R$^-$	Ionicity	Avail. (pm)
FeBr$_2$ [WYC 63a]	374.0	617.0	0.311	27%	+8.4
FeI$_2$ [WYC 63a]	404.0	675.0	0.277	16%	+6.9
MgI$_2$ [WYC 63a]	414.0	688.0	0.327	37%	+2.5
MgBr$_2$ [WYC 63a]	381.0	626.0	0.367	49%	+2.0
ZnI$_2$ [WYC 63a]	425.0	654.0	0.336	24%	+0.9
CoBr$_2$ [THO 40]	368.0	612.0	0.332	25%	+0.8
ZrS$_2$ [WYC 63a]	363.0	585.0	0.391	33%	−0.4
TiS$_2$ [TRO 81]	341.3	571.7	0.332	24%	−1.6
CoI$_2$ [WYC 63a]	396.0	665.0	0.295	14%	−2.3
TiSe$_2$ [EHR 48]	354.8	599.8	0.308	24%	−5.1
PtS$_2$ [GRO 60]	354.3	503.9	0.342	4%	−6.8
MnI$_2$ [WYC 63a]	416.0	682.0	0.377	27%	−8.5
SnS$_2$ [FIL 16]	354.6	525.7	0.375	10%	−9.7
TiTe$_2$ [ARN 81]	377.7	649.8	0.276	8%	−10.1
PtSe$_2$ [FUR 65]	372.8	503.1	0.318	4%	−11.7
IrTe$_2$ [JOB 01]	399.1	547.1	0.285	0%	−16.0
PtTe$_2$ [FUR 65]	402.6	522.1	0.285	0%	−17.4
NiTe$_2$ [WYC 63a]	386.1	529.7	0.312	1%	−30.7
PdTe$_2$ [WYC 63a]	403.7	512.6	0.389	0%	−41.0

Table 3.23. *Several compounds of CdI$_2$ type (R$^+$/R$^-$ < 0.414)*

Table 3.24 lists compounds whose steric criterion corresponds to coordination 6 and whose ionicity is higher than that of the previous series. The hard sphere model can be validated for the compounds with appropriate availability, and negative availability can be observed when covalence increases. The positive availability of $FeBr_2$ can be linked to anionic availability, which is negative (−18.0 pm) and prevents the ions from moving closer.

The compounds with negative availability have also negative anionic availability and weak ionicity, and an all the stronger covalence of their bond can be proposed.

Compounds (MX_2)	a	c	R^+/R^-	Ionicity	Availability (pm)
PtO_2 [HOE 71]	310.0	416.1	0.450	34%	+4.0
$Co(OH)_2$ [WYC 63a]	317.3	464.0	0.439	27%	+3.8
$Mg(OH)_2$ [DES 96]	314.8	477.9	0.486	51%	−2.5
$Ni(OH)_2$ [WYC 63b]	311.7	459.5	0.466	26%	−3.5
$Mn(OH)_2$ [WYC 63a]	334.0	468.0	0.561	41%	−5.4
VCl_2 [VIL 59]	360.1	583.5	0.436	44%	−6.0
$Ca(OH)_2$ [WYC 63a]	358.4	489.6	0.676	63%	−7.6
CaI_2 [WYC 63a]	448.0	696.0	0.455	50%	−8.3
$MnBr_2$ [WYC 63a]	382.0	619.0	0.423	39%	−9.6
YbI_2 [DOE 39]	448.0	696.0	0.464	41%	−10.3
$TiCl_2$ [WYC 63a]	356.1	587.5	0.475	48%	−14.3
CdI_2 [DE 69]	424.0	685.9	0.432	21%	−16.1
$TiBr_2$ [EHR 61]	362.9	649.2	0.439	40%	−17.0
PbI_2 [WYC 63a]	455.5	697.7	0.541	17%	−23.4

Table 3.24. *Several compounds of CdI_2 type ($0.414 < R^+/R^- < 0.732$)*

The CdI_2 type contains a majority of strongly covalent compounds, which validate the hard sphere model. A negative anionic availability can also be noted, which is associated to the negative availability of cations, indicating a strong covalence.

3.4.5. Study of cadmium chloride

3.4.5.1. Description of $CdCl_2$ unit cell based on ion lattices

The compounds of cadmium chloride type ($CdCl_2$) are much less numerous than those of CdI_2 type, but they are another alternative of packing close to the previous one. Moreover, several compounds have both types ($CdBr_2$, $FeCl_2$, $MnCl_2$ and ZnI_2).

The unit cell is rhombohedral (1 and 2) and by indicating several atoms outside the unit cell (see Figure 3.72), the ABC packing can be visualized as materialized by two cuboctahedra (3) of anions and a second packing of cations of ABC type but with very distant planes (4). The distance between two planes of anions of c/6 and between two planes of cations of c/3 (3 and 4) is visualized.

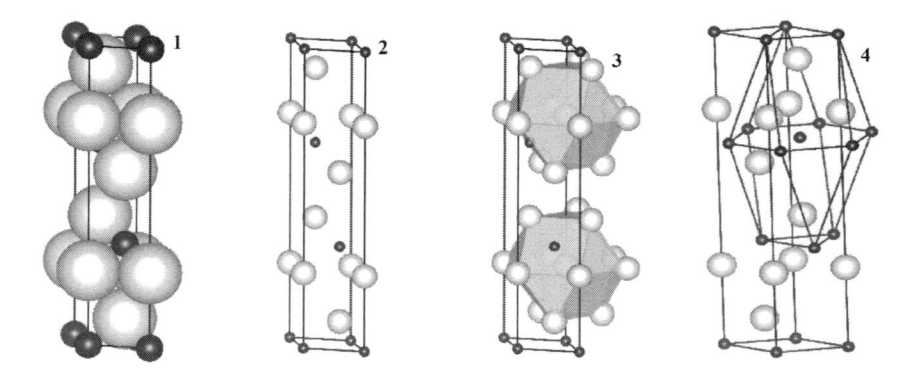

Figure 3.72. *Representation of coordination polyhedra of CdI_2. For a color version of this figure, see www.iste.co.uk/valls/inorganic.zip*

3.4.5.2. Description of $CdCl_2$ unit cell based on coordination polyhedra

In the $CdCl_2$ unit cell (1 and 2), cations are in octahedra (3) and the anions, similarly to the case of CdI_2 (see Figure 3.73), are in coordination 3 in the form of pyramid with triangular base (4). This arrangement satisfies the electrostatic valence rule and shows the anisotropy of the environment of ions and therefore the covalent character of bonds.

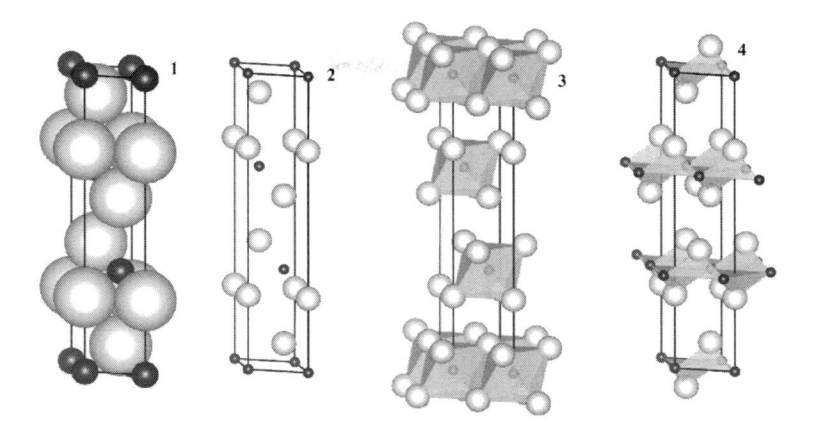

Figure 3.73. *Representation of coordination polyhedra between CdI_2 ions. For a color version of this figure, see www.iste.co.uk/valls/inorganic.zip*

3.4.5.3. *Calculation of various values of $CdCl_2$ unit cell*

Availability can be obtained by calculating the distance between the two closest possible opposite ions, similarly to CdI_2.

On the basis of ABC triangles (1 and 2) represented in Figure 3.74, it can be noted that AC is equal to the coordinate along c of atom A minus the coordinate along c of atom B, which leads to $\frac{c}{3}-\frac{c}{4}$ or $\frac{c}{12}$.

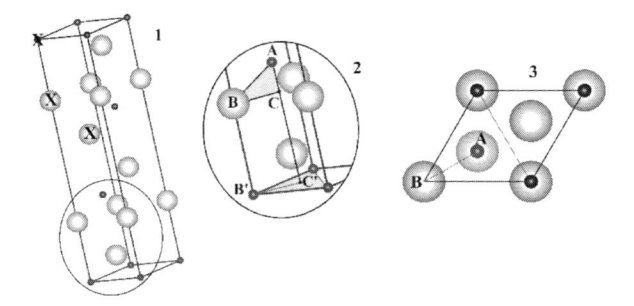

Figure 3.74. *Study of CdI_2 unit cell. For a color version of this figure, see www.iste.co.uk/valls/inorganic.zip*

AB can be readily calculated on the basis of a top view of the unit cell (3) since it is equal to 2/3 of the altitude of the equilateral triangle or $\frac{2}{3} a\frac{\sqrt{3}}{2}$ or $\frac{a}{\sqrt{3}}$.

The application of the Pythagorean theorem yields the distance between anion and cation from which the anion radius is subtracted and spatial availability results:

$$\text{Availability} = \sqrt{\left(\tfrac{c}{12}\right)^2 + \left(\tfrac{a}{\sqrt{3}}\right)^2} - R^- \qquad [3.2]$$

3.4.5.4. Compounds of CdCl₂ type

Compounds of $CdCl_2$ type are halides of ions with 2 charges and are much less numerous than the compounds of CdI_2 type.

Only two compounds have a correct steric criterion, namely $CdCl_2$ and $CdBr_2$, but their availability is strongly negative. Their ionicities are 42 and 33%, respectively; therefore, they can be assigned a part of covalence in order to explain availabilities.

For all the other compounds, the steric criterion is below the expected value, but the passage to coordination 4 cannot be envisaged, in contrast to the CdI_2 type.

These compounds validate the hard sphere model and it can be noted that, contrary to the CdI_2 type, anionic availability is correct. Only the compounds of weak ionicity $NiBr_2$ and NiI_2 have negative availability (Avail.) and they can be assigned a strong covalent character (see Table 3.25).

Compounds (MX₂)	R⁺; R⁻	a	c	R⁺/R⁻	Ionicity	Avail. (pm)
CdCl₂ [WYC 63a]	95; 181	385.0	1746	0.525	42%	−10.3
CdBr₂ [WYC 63a]	95; 196	395.0	1867	0.485	33%	−14.9
MgCl₂ [BUS 70]	72; 181	359.5	1759	0.398	57%	+1.1
MnCl₂ [WYC 63a]	83; 181	368.6	1747	0.459	48%	−6.2
ZnBr₂ [WYC 63a]	74; 196	392.0	1873	0.376	35%	+4.9
CoCl₂ [WYC 63b]	65; 181	354.4	1743	0.359	34%	+4.9
NiCl₂ [WYC 63a]	69; 181	354.3	1734	0.381	32%	+0.4
NiBr₂ [WYC 63a]	69; 196	370.8	1830	0.352	24%	−2.2
ZnI₂ [WYC 63a]	74; 220	425.0	2150	0.336	23%	+9.8
NiI₂ [WYC 63a]	69; 220	389.2	1963	0.314	13%	−11.0

Table 3.25. *Several compounds of CdCl₂ type*

Knowledge of Ionic Crystals 215

NOTE 3.22.– There are important similarities between the two unit cells, but CdCl₂ has (1) an ABC packing (cuboctahedron), whereas CdI₂ has (2) a hexagonal close-packed structure (AB with anti-cuboctahedra).

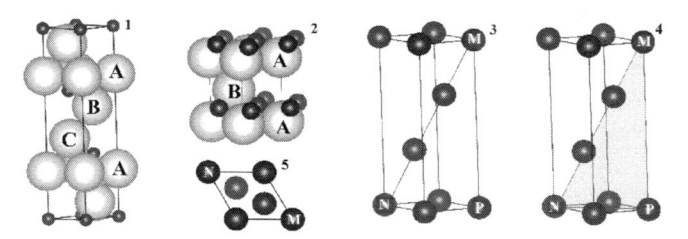

For a color version of this figure, see www.iste.co.uk/valls/inorganic.zip

These are close neighboring packings, but the distance between cations is more significant than in the case of CdCl2.

In MNP triangle (3 and 4), the side MP is equal to the lattice parameter c, and NP is twice the altitude of CDE triangle or $2\frac{a\sqrt{3}}{2}$ or $a\sqrt{3}$.

The application of the Pythagorean theorem to MNP triangle yields the side:

$$MN = \sqrt{\left(a\sqrt{3}\right)^2 + c^2}$$

If ZnI2 is considered to present these two types, the distance between cations is 757.5 pm in the CdCl₂ type and 654 pm in the CdI₂ type. The approach of cations for CdI₂ can be thought destabilizing, and it explains the much less important number of compounds in this crystalline type.

3.5. Review of characteristics of binary structures

3.5.1. *Crystalline characteristics*

Table 3.26 lists the characteristics of typical compounds, coordination of ions, the space group (SG), the number of formula units (Z), the volume attached to the unit cell (Volume), the indication of topologically identical ions (Topol.) and the formula stoichiometry (Stoic.).

Compounds	Coordination	SG	Z	Volume	Topol.	Stoic.
CsCl	8:8	P m–3m	1	cube	X	1/1
NaCl	6:6	F m–3m	4	cube	X	1/1
ZnSs	4:4	F –43m	2	cube	X	1/1
CaF_2	8:4	F m–3m	4	cube		1/2
Li_2O	4:8	F m–3m	4	cube		2/1
ZnSw	4:4	P 6_3mc	2	hexagon	X	1/1
NiAs	6:6pt	P 6_3/mmc	2	hexagon		1/1
$CdCl_2$	6:3py	R –3m	1	hexagon		1/2
CdI_2	6:3py	P –3m1	1	hexagon		1/2
TiO_2	6:3tr	P 4_2 / mnm	2	tetragonal prism		1/2

Table 3.26. *Several characteristics of the described typical compounds*

The coordinations should be detailed, as the same number may denote several polyhedra, "3 tri" indicates plane coordination at 120° while "3 py" indicates one ion surrounded by three opposite ions, the set forming a pyramid with triangular base. Coordination 4 is only tetrahedral (4 tetra), whereas coordination 6 can be octahedral (6 oct), in which the ions can form a trigonal prism (6 pris). Finally, in cubic coordination 8, the ions form a cube with an oppositely charged ion at its center. Coordinations 3 or 6 have two possibilities indicated by the concerned atom being marked by underline or an asterisk (see Table 3.27).

Compounds	3 <u>tri</u> and 3 py*	4 tetra	6 <u>oct</u> and 6 pris*	8 cube
MX	–	ZnS s and w	<u>NaCl</u>, NiAs*	CsCl
MX_2	CdI_2*, $CdCl_2$*, Ti<u>O</u>$_2$	Li<u>O</u>$_2$, Ca<u>F</u>$_2$	<u>Ti</u>O_2, <u>Cd</u>I_2, <u>Cd</u>Cl_2	<u>Ca</u>F_2, <u>Li</u>O_2

Table 3.27. *Coordination of ions in typical crystals*

3.5.2. Characteristics of availability

The correlation between ionicity and availability (see Figure 3.75) for all the ionic covalent compounds (except intermetallic compounds) is described.

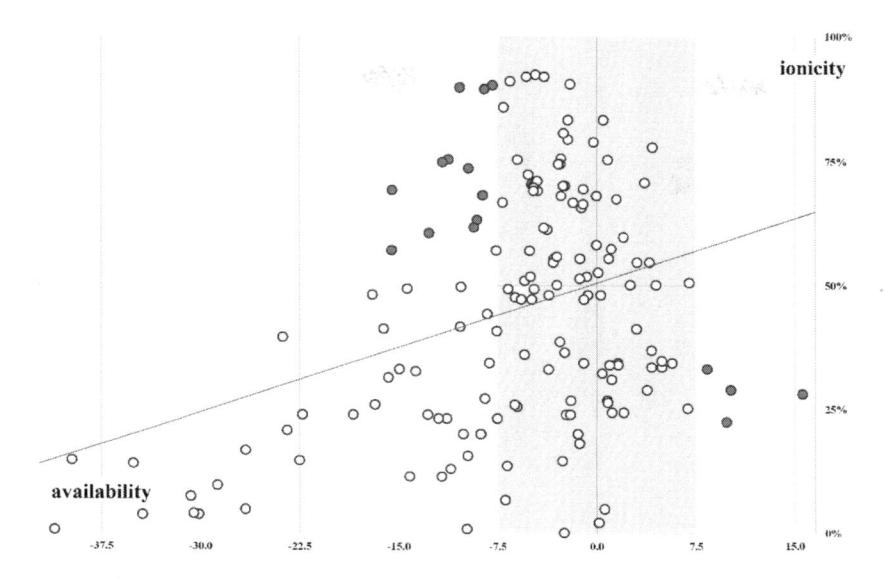

Figure 3.75. *Correlation between ionicity and availability for all the ionic covalent compounds. For a color version of this figure, see www.iste.co.uk/valls/inorganic.zip*

Outside the tolerance interval, there is a visible tendency toward decreased availability when ionicity diminishes, despite a certain dispersion of values. A total of 16 (colored) compounds with positive criterion and weak ionicity or the reverse can be observed, but these are considered exceptions.

3.5.3. *Characteristics of the unit cells*

The unit cells whose basic volume is cube are CsCl (1), NaCl (2), ZnS sphalerite (3), CaF_2 (4) and Li_2O (5). They all have ionic covalent bonds, knowing that for ZnS sphalerite type, all the compounds have weak ionicity (see Figure 3.76).

The hard sphere model fits their description, and deviations from model are generally encountered for compounds of weak iconicity, for which covalence brings ions closer and renders their availability negative.

The three first unit cells are of MX type with topologically identical ions. The last two are of MX_2 type and the coordination of cations is different from that of anions ((4:8) and (8:4), respectively).

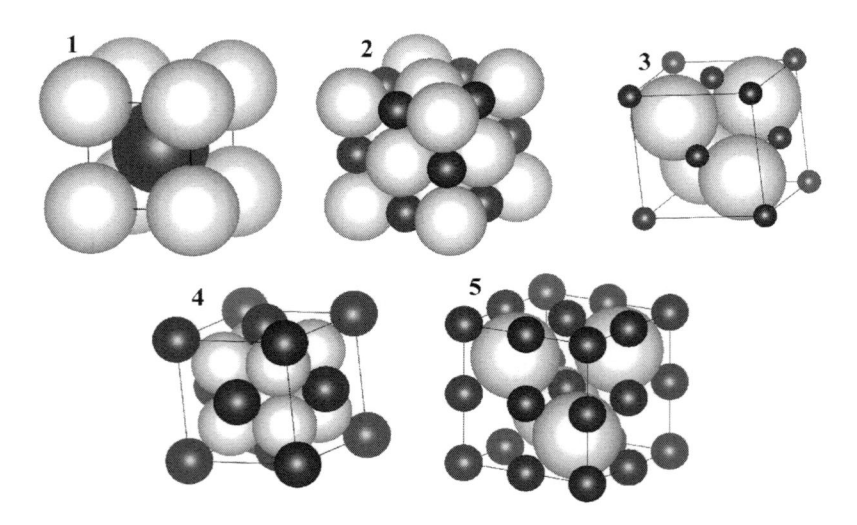

Figure 3.76. *Compounds with cubic unit cell (ClCs, ClNa, ZnS, CaF$_2$ and Li$_2$O)*

The unit cells having hexagon as basic volume are ZnS wurtzite (1), NiAs (2), CdI$_2$ (3) and CdCl$_2$ (4) and they all have a mainly covalent bond (see Figure 3.77).

The first two are of MX type and the last two are of MX$_2$ type, and the ions are topologically identical only for ZnS.

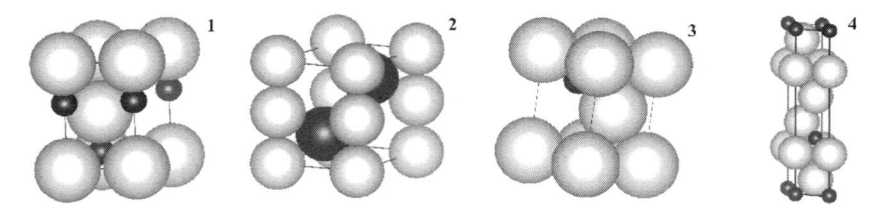

Figure 3.77. *Compounds with hexagonal unit cell (ZnS wurtzite, NiAs, CdI$_2$ and CdCl$_2$)*

The hard sphere model can describe these compounds knowing that they are strongly covalent, except for NiAs, which has partly metallic and partly covalent character.

3.5.4. *Characteristics of the families of compounds*

3.5.4.1. *Halides*

The family-based analysis of compounds shows that alkaline elements and the silver ion with one charge yield compounds of NaCl type, with the exception of large Tl and Cs ions, which yield CsCl type (see Figure 3.78).

It can be observed that under very high pressure, there is a shift of RbCl, NaCl and AgI toward the CsCl type.

The alkaline-earth compounds (except for magnesium ion) yield CaF_2 type and then an orthorhombic (ortho) type, provided as illustration.

The elements in the block with 2 charges as well as the magnesium-ion yield compounds of TiO_2 type for fluorides (due to its electronegativity that is above that of the other elements in its column) and then layered compounds, of CdI_2 and $CdCl_2$ types.

Halides of copper ions with 1 charge are all of ZnS sphalerite type.

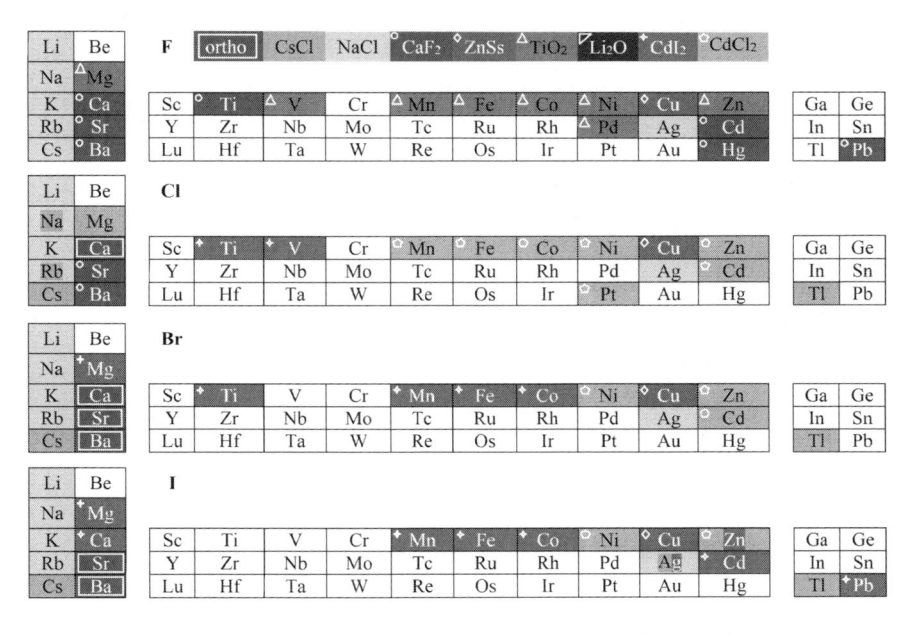

Figure 3.78. *Positioning of halides in the classification. For a color version of this figure, see www.iste.co.uk/valls/inorganic.zip*

3.5.4.2. Chalcogenides

Similarly to halides, only the first element (oxygen of higher electronegativity compared to other elements in its column) yields in combination with elements of d-block, depending on formula stoichiometry, chalcogenides of TiO_2 type or compounds of NaCl type (see Figure 3.79).

Several multi-type compounds, such as titanium, manganese, platinum, silver, zinc or lead, can be observed. The presence of several types is moreover common for column 16.

It is worth noting the presence of NiAs type, which is absent among more ionic halides.

The alkaline compounds yield the Li_2O type, and the alkaline-earth compounds yield the NaCl type, except for beryllium, which is of ZnS sphalerite or wurtzite type.

Elements of d-block (except for oxides) yield a large variety of layered compounds of CdI_2, $CdCl_2$ and NiAs type as well as compounds of NaCl type for the less electronegative or ZnS sphalerite and wurtzite.

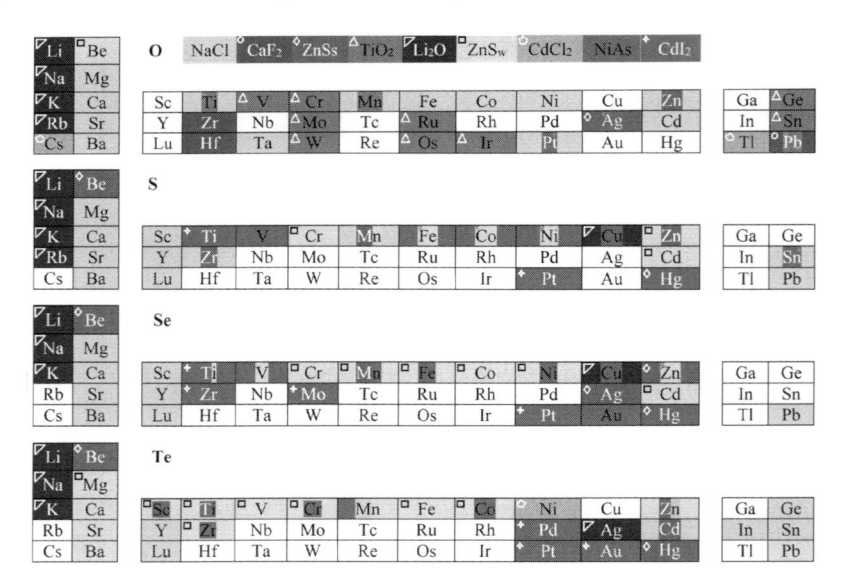

Figure 3.79. *Positioning of chalcogenides in the periodic table. For a color version of this figure, see www.iste.co.uk/valls/inorganic.zip*

3.6. Geometry of ternary crystals of AB_nO_m type

3.6.1. *Study of $SrTiO_3$ perovskite*

Perovskites are the first typical compounds described that simultaneously involve (at least) two types of cations for only one type of anion.

The perovskite structure derives its name from calcium titanium oxide ($CaTiO_3$), but we propose strontium titanium oxide ($SrTiO_3$), which better represents the dominant space group P m-3m ($CaTiO_3$ belonging to orthorhombic space group [KNI 09] P bnm).

Perovskites derive from oxides of AO and BO_2 type, which are submitted to reaction at high temperature and yield ABO_3. A and B cations have very different ionic radii (atoms A are larger than atoms B) and X an anion (generally oxygen and sometimes fluorine or chlorine).

The unit cell is cubic and therefore in order to analyze various characteristics, the following three ions can be successively placed at the vertices of the cube (see Figure 3.80): strontium in $SrTiO_3$ (1), titanium in $TiSrO_3$ (2) and oxygen in O_3TiSr (3).

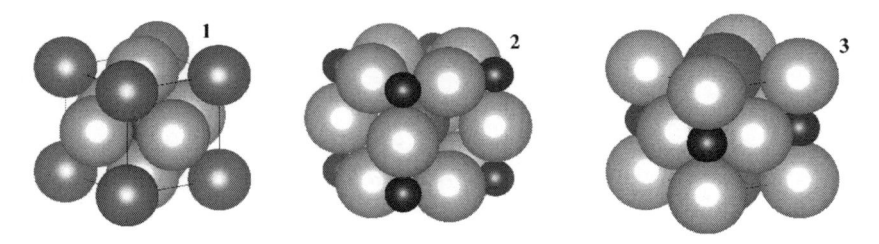

Figure 3.80. *Various representations of the unit cell of perovskites. For a color version of this figure, see www.iste.co.uk/valls/inorganic.zip*

There are several thousand compounds of perovskite type, but the focus here will be limited to those of formula $A_1B_1O_3$, which can be classified, for example, on the basis of oxidation numbers of the two constituting cations and of symmetry groups.

Several examples have been chosen (see Figure 3.82) in each space group and a larger number, for the most common P m-3m group. Mainly oxides, several examples of fluorides and a chloride are proposed.

3.6.1.1. Description of SrTiO₃ unit cell based on ion lattices

The unit cell can be divided into three ion lattices that interlink and form a crystal (see Figure 3.81): a cubic P lattice of strontium ions that form octahedra (1) linked by the sides and an identical lattice shifted by the semi-diagonal of the cube for the titanium ions (2), whereas the oxygen ions form a tetragonal lattice I (3) for which two unit cells have been represented.

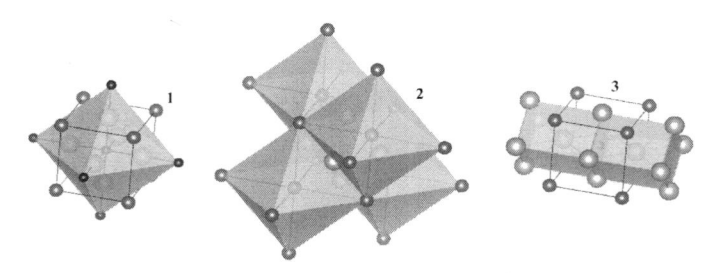

Figure 3.81. *Coordination polyhedra in a perovskite. For a color version of this figure, see www.iste.co.uk/valls/inorganic.zip*

			space group			
II-IV	PbTiO₃	◁	P m -3 m	▷	AlLaO₃	III-III
II-IV	CaTiO₃	◁	P m -3 m	▷	LaFeO₃	III-III
II-IV	EuTiO₃	◁	P m -3 m			
II-IV	SrTiO₃	◁	P m -3 m	▷	NaMgF₃	I-II
II-IV	BaIrO₃	◁	P m -3 m	▷	KCoF₃	I-II
II-IV	SrGeO₃	◁	P m -3 m	▷	RbCaF₃	I-II
II-IV	CaSiO₃	◁	P m -3 m	▷	KMgF₃	I-II
II-IV	BaSnO₃	◁	P m -3 m	▷	CsPbCl₃	I-II
II-IV	BaZrO₃	◁	P m -3 m			
II-IV	SrSnO₃	◁	P m -3 m	▷	NaNbO₃	I-V
II-IV	SrZrO₃	◁	P m -3 m			
II-IV	SrTa₀.₅V₀.₅O₃	◁	P m -3 m			
II-IV	BaCoO₂.₂₂	◁	P m -3 m			
II-IV	BaBiO₃	◁	F m -3 m	▷	K₂SiF₆	I-IV
II-IV	PbTiO₃	◁	P 4mm			
II-IV	BaTiO₃	◁	P 4mm			
II-IV	BaSiO₃	◁	P 63/m m c	▷	RbZnF₃	I-II
II-IV	BaMnO₃	◁	P 63/m m c			
II-IV	MgSiO₃	◁	P b n m			
II-IV	CaTiO₃	◁	P b n m			
II-IV	MnNdO₃	◁	P n m a	▷	BiInO₃	III-III
II-IV	CaVO₃	◁	P n m a	▷	LaMnO₃	III-III
			P n m a	▷	NaMgF₃	I-II
			R -3 c :H	▷	LaNiO₃	III-III
			R -3 c :H	▷	AlLaO₃	III-III

Figure 3.82. *Type and space group of various perovskites. For a color version of this figure, see www.iste.co.uk/valls/inorganic.zip*

3.6.1.2. Description of the $SrTiO_3$ unit cell based on coordination polyhedra

The ideal structure (see Figure 3.83) is simple cubic and can be represented in the form ABX_3 (1 and 2) or BAX_3 (3 and 4), in which ions A and B are not surrounded by the same number of oxygen atoms. Indeed, when cations are permuted, the arrangement of oxygen ions is not the same in the unit cell.

Cations B are surrounded by six anions (see Figure 3.83) that form an octahedron (1 and 2), and cations A are surrounded by 12 anions that form a cuboctahedron (3 and 4).

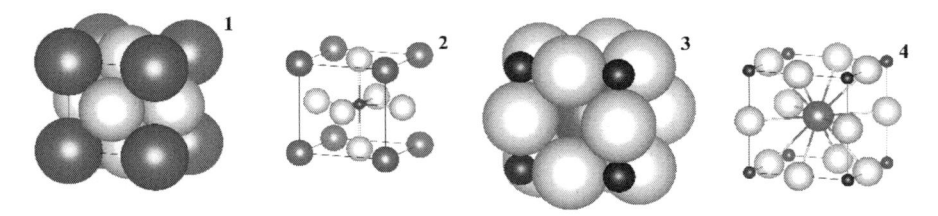

Figure 3.83. *Unit cell corresponding to $SrTiO_3$ and $TiSrO_3$. For a color version of this figure, see www.iste.co.uk/valls/inorganic.zip*

The unit cell is cubic P (see Figure 3.84) and contains a formula unit cell that can be defined on the basis of three representations of the unit cell: $TiSrO_3$ (1), $SrTiO_3$ (2) and the last in which no atom occupies the vertices of the cube (3).

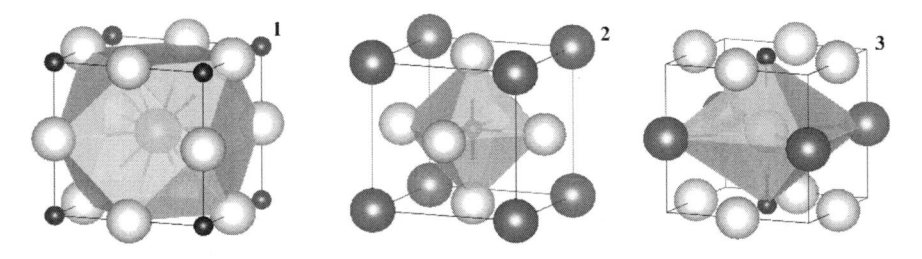

Figure 3.84. *Coordination polyhedra of ions in a perovskite. For a color version of this figure, see www.iste.co.uk/valls/inorganic.zip*

The anions X are in mixed octahedral coordination (3) surrounded by four cations A (square plane coordination) and two cations B that complete the cuboctahedron (linear coordination).

3.6.1.3. *Calculation of various values of the SrTiO₃ unit cell*

In the first structure, there are $8 \times 1/8$ or one strontium ion, one (non-visible) titanium ion at the center and $6 \times 1/2$ or three oxygen ions. In the second one, there is $8 \times 1/8$ or one titanium ion, one strontium ion at the center and $12 \times 1/4$ or three oxygen ions. Finally, the third yields $4 \times 1/4$ titanium ion, $2 \times 1/2$ strontium ion and $8 \times 1/8 + 4 \times 1/2$ or three oxygen ions.

The lattice parameter is equal to 390.5 pm, and the ion radii have the following values: 144 pm for the strontium ion, 61 pm for the titanium ion and 140 pm for the oxygen ion.

Figure 3.84 (1) shows that $(a/2 - R^-)$ can be used to calculate the availability of titanium, which is −5.7 pm.

Similarly, (2) yields $(\frac{a\sqrt{2}}{2} - R_O)$, or −7.9 pm, which is the availability of strontium.

The calculation of ionicity yields 80% for the Sr–O bond and 62% for Ti–O bond; the titanium ion being polarizing, it leads to a ionic covalent bond that fixes parameter a and actually reduces the availability of strontium.

Reality is never simple, and the interaction between sets of ions can be imagined, in order to obtain the most stable equilibrium. An availability of −17.2 pm had been observed for TiO (NaCl type), which is impossible for perovskite. To obtain its place, the strontium ion cannot allow such significant approach between titanium and oxygen, and the availability of titanium is within the tolerance interval, with a value of −5.7 pm!

The same phenomenon can be noted for $SrGeO_3$ (see Table 3.28), but the germanium ion, more polarizing, imposes a short-distance Ge–O (availability of −23.1 pm) that reduces the availability of strontium (−15.5 pm).

3.6.1.4. Compounds of $SrTiO_3$ type

Perovskites have a wide variety of properties depending on the choice of elements A and B. This variety comprises compounds that have, for example, ferroelectric properties ($BaTiO_3$ [KWE 93]), ferromagnetic properties ($YTiO_3$ [HES 97]), etc.

> NOTE 3.23.– Many properties of perovskites result, for example, from combinations of elements ($La_{0.1}Na_{0.1}Sr_{0.8}TiO_3$, $KCu_{0.3}Mg_{0.7}F_3$ [MIT 00]), from oxygen non-stoichiometry ($Ga_{0.8}LaMg_{0.2}O_{2.9}$ [KAJ 03]) or from overall stoichiometry ($Ca_{0.96}Mn_{0.32}Si_{0.64}O_{2.56}$ [GOL 43]).

It is generally not interesting to mention compounds that have not been described, but in this case, it serves to evidence the infinite possibilities offered by perovskites, which are too specialized within the framework of this book.

Materials with a perovskite structure are used in sensors, memory devices (RAM), amplifiers, fuel cells, optoelectronic devices, such as high-temperature superconductors, etc.

Examples of perovskite of ABO_3 formula of P m-3m space group have been chosen, this being the most important and generally proposed type to qualify these compounds (see Table 3.28).

The compounds are classified according to the availability of the smallest cation (Avail. B) in three groups depending on whether the anion is oxygen, fluorine or chlorine ion.

In contrast with the previous types, the steric criterion cannot be readily interpreted, since for perovskites, the two cations play a role in the equilibrium of the crystal and the two ratios are never simultaneously respected.

Compounds ABX$_3$	a	R^+_A/R$^-$	Avail. A	R^+_B/R$^-$	Avail. B
LaFeO$_3$ [KOE 57]	392.6	0.971	+1.6	0.393	+1.3
BaIrO$_3$ [CHE 09]	406.1	1.150	−13.8	0.450	+0.1
CaSiO$_3$ [KAN 91]	356.4	0.957	−22.0	0.286	−1.8
NaNbO$_3$ [WYC 31]	389.0	1.045	+2.1	0.481	−2.5
PbTiO$_3$ [YOS 16]	396.4	1.064	−8.7	0.436	−2.8
BaZrO$_3$ [MEG 46]	418.2	1.150	−5.3	0.514	−2.9
BaSnO$_3$ [MEG 46]	410.9	1.150	−10.5	0.493	−3.6
SrSnO$_3$ [MEG 46]	409.3	1.029	+5.4	0.493	−4.3
LaAlO$_3$ [NAK 05]	379.1	0.971	−7.9	0.386	−4.4
SrTiO$_3$ [MIT 00]	390.5	1.029	−7.9	0.436	−5.7
EuTiO$_3$ [BRO 53]	390.5	0.964	+1.1	0.436	−5.8
SrZrO$_3$ [MEG 46]	408.9	1.029	+5.1	0.514	−7.6
CaTiO$_3$ [BAR 25]	379.5	0.957	−5.7	0.436	−11.3
SrGeO$_3$ [NAK 15]	379.8	1.029	−15.5	0.521	−23.1
KCoF$_3$ [KNO 61]	407.1	1.233	−9.1	0.489	+5.6
KMgF$_3$ [BUR 96]	398.6	1.233	−15.2	0.541	−5.7
RbCaF$_3$ [HUT 81]	444.5	1.293	+9.3	0.752	−10.8
NaMgF$_3$ [CHE 05]	387.6	1.045	+2.1	0.541	−11.2
CsPbCl$_3$ [MOL 58]	560.5	1.039	+27.3	0.431	+21.3

Table 3.28. *Several perovskites of P m-3m space group*

Their calculation shows that the compounds respect the value corresponding to coordination 6 for the large cation (R^+_A/R$^-$) but only some of them (LaAlO$_3$, CaSiO$_3$ and LaFeO$_3$) obey the coordination 4 for the other cation (R^+_B/R$^-$).

Availability is within the tolerance interval for ion B (except for the most covalent ones in $CaTiO_3$ and $SrGeO_3$) and it is correct for half of the ions A (Avail. A), knowing that the other half has covalent character. The hard sphere model can be validated by indicating part of covalence in many cases for one cation ($SrZrO_3$), the other cation ($BaIrO_3$) or both cations ($SrGeO_3$). In the latter case, availability is also negative for the anions (−11.5 pm for $SrGeO_3$ and $CaTiO_3$) and goes in the sense of the part of covalence.

The two cations contribute to the equilibrium of the crystal and have to meet demands that are sometimes contradictory. For example, $RbCaF_3$ has an availability of −10.8 pm for A and of +9.3 pm for B, reducing it for B is difficult, as it requires a reduction also for A, while it is already negative. It can be noted that availabilities of $KCoF_3$ have opposite signs compared to those of $RbCaF_3$, but the same contradiction in terms of evolution.

The interdependence of the two cations in these crystals is confirmed, since 75% of the compounds are concerned by this phenomenon (there are still 25% that satisfy both availabilities).

These interactions in perovskites can lead to a deformation of crystals and induce a change in symmetry (space groups P bnm, P 4 mm, P nma, R -3c, etc. can be observed), which qualifies them as choice compounds in terms of varied properties and experimental research full of opportunities.

3.6.2. *Study of $MgAl_2O_4$ spinel*

Similarly to perovskites, spinels involve (at least) two types of cations for only one type of anion.

Spinel structure is almost exclusively of space group F d-3m, and in order to describe it the compound $MgAl_2O_4$ [ISH 82] is used.

Spinels derive from oxides of AX and B_2X_3 type that are brought to react at high temperatures and yield AB_2X_4. Cations A and B have different charges and coordinations and X is an anion (generally oxygen or sulfur).

There are several thousand compounds of spinel type, but the focus will be limited here to those of formula $A_1B_2X_4$ that can be classified, similar to perovskites, on the basis of oxidation numbers of the two constituting cations, and are divided into oxides and sulfides, as the only space group present is F d-3m. Examples have been chosen (see Figure 3.85) for various oxidation numbers, most of them relating to group II-III, which is the most common.

oxides	F d -3 m	sulfides
$ZnCr_2O_4$ ◁	II-III ▷	$MgYb_2S_4$
$CuAl_2O_4$ ◁	II-III ▷	$MnYb_2S_4$
$PtMg_2O_4$ ◁	II-III ▷	$FeYb_2S_4$
$CuCr_2O_4$ ◁	II-III ▷	$MgLu_2S_4$
MnV_2O_4 ◁	II-III ▷	$FeLu_2S_4$
ZnV_2O_4 ◁	II-III ▷	$MgSc_2S_4$
FeV_2O_4 ◁	II-III ▷	$FeSc_2S_4$
CdV_2O_4 ◁	II-III ▷	$MnLu_2S_4$
$MgAl_2O_4$ ◁	II-III ▷	$CuCo_2S_4$
MgV_2O_4 ◁	II-III ▷	$MnIn_2S_4$
$MgGa_2O_4$ ◁	II-III ▷	$CoIn_2S_4$
VZn_2O_4 ◁	IV-II	
VMn_2O_4 ◁	IV-II	
$SnZn_2O_4$ ◁	IV-II	
VCo_2O_4 ◁	IV-II	
VMg_2O_4 ◁	IV-II	

Figure 3.85. *Type and space group of various spinels. For a color version of this figure, see www.iste.co.uk/valls/inorganic.zip*

NOTE 3.24.– This section presents only the most simple compounds and ignores compounds with three or more cations ($CrMnNiO_4$ [REN 72]), such as the compounds with balanced fractional stoichiometry ($Cd_{1.2}SnZn_{0.8}O_4$ [CHO 68]) or non-balanced fractional stoichiometry ($Cd_{1.333}Sn_{1.333}O_4$ [LEV 70]), such as oxygen sub-stoichiometric compounds ($CuNdO_{2.93}$ [CHE 95]), as the more complex compounds ($Co_{1.041}Li_{1.959}Ti_{3.001}O_8$ [JOV 03] or $Ba_{0.67}Bi_{0.68}K_{0.31}Sm_{0.34}O_3$ [BEZ 93]).

Similar to perovskites, the objective is to highlight the infinite possibilities offered by spinels and to inspire an interest to go beyond the scope of this book.

3.6.2.1. *Description of MgAl2O4 unit cell based on ion lattices*

The spinel structure can be described (see Figure 3.86) as a diamond packing of magnesium ions (1) interlinked with a cubic F packing of oxygen ions (2 and 3) and the aluminum cations form five tetrahedra inside the unit cell (4).

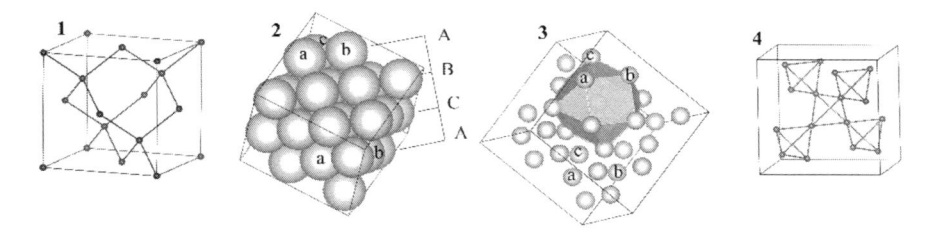

Figure 3.86. *Packings in spinels*

The best way to describe the spinel structure is to divide the cube corresponding to the unit cell into eight cubes (see Figure 3.87).

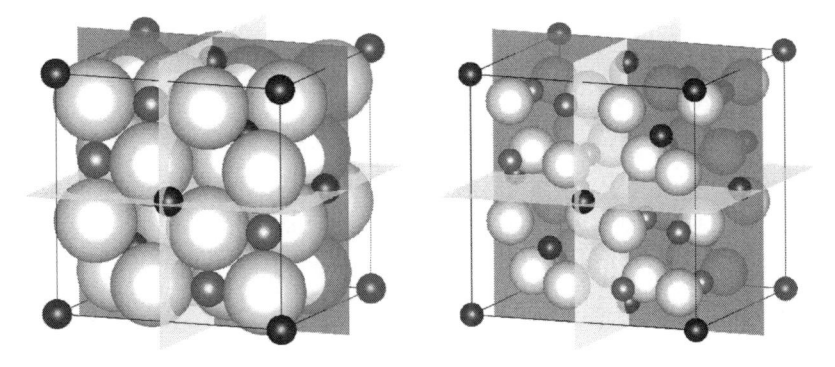

Figure 3.87. *Real unit cell and its representation with ions reduced by one spinel. For a color version of this figure, see www.iste.co.uk/valls/inorganic.zip*

An alternation is observed in the three directions of small cubes containing either a tetrahedron of oxygen ions (1) or cubes that alternate aluminum and oxygen ions (2).

The complete unit cell (4) has been represented (see Figure 3.88) as well as the set of ions that are fully or partially present in the first type of small cube (5), as well as in the second type of small cube (6).

Magnesium ions are distributed according to a diamond packing (3).

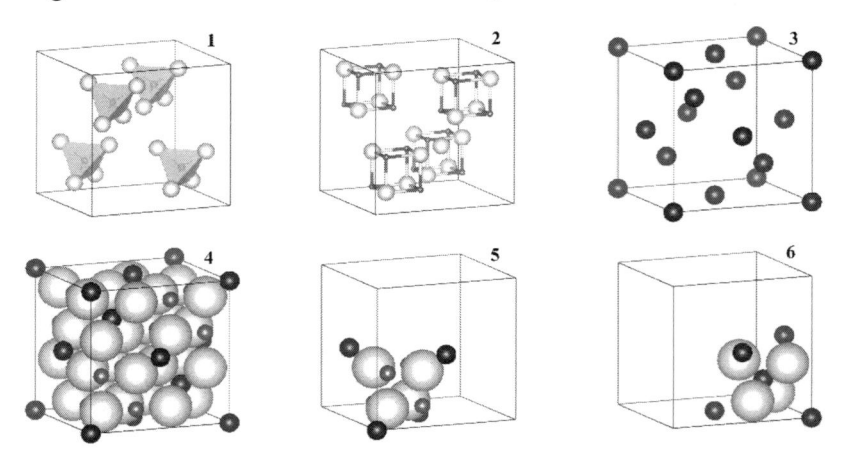

Figure 3.88. *Various representations of spinel unit cell. For a color version of this figure, see www.iste.co.uk/valls/inorganic.zip*

3.6.2.2. *Description of MgAl$_2$O$_4$ unit cell based on coordination polyhedra*

The spinel structure can be described as a cubic F packing (1 and 2) of oxygen ions (see Figure 3.89) featuring cuboctahedra (3).

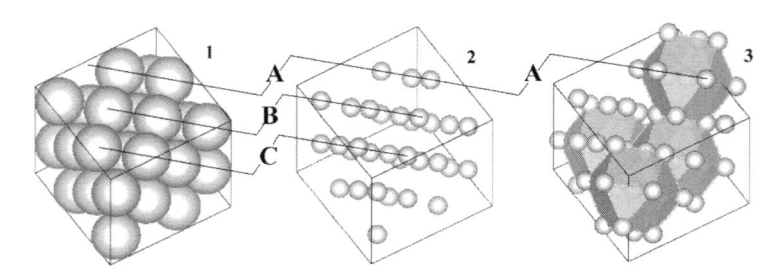

Figure 3.89. *Packing of oxygen ions in the spinels*

It is a cubic F packing of anions (ABC packing along direction [111] of the spinel unit cell), in which 50% of the octahedra sites are occupied by a cation and 1/8 of the tetrahedral sites are occupied by the second cation.

Each unit cell contains eight formula units and therefore 32 anions associated to 32 octahedral sites and 64 tetrahedral sites.

Figure 3.90 shows representations of octahedral polyhedra of oxygen ions around aluminum ions (1) and tetrahedra of oxygen ions around magnesium ions (3), as well as the complete unit cell (2). The coordination of oxygen ions is double, with three aluminum ions and one magnesium ion forming a distorted tetrahedron (4 and 5).

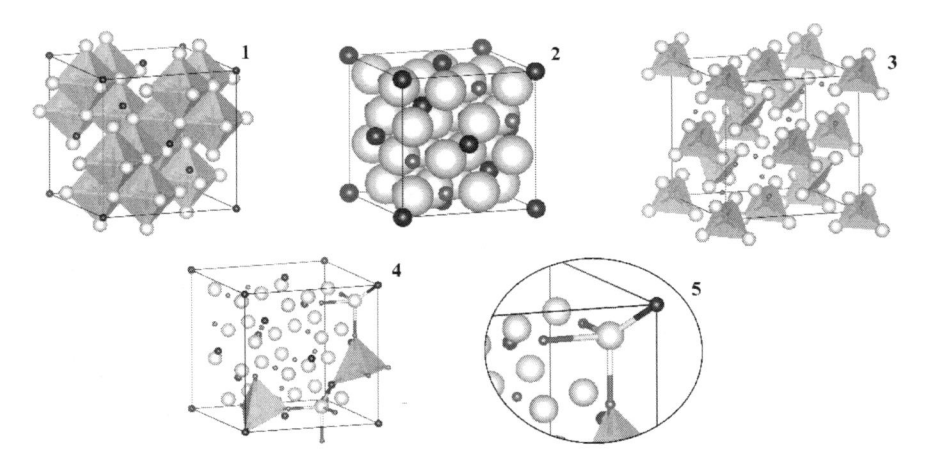

Figure 3.90. *Coordination polyhedra of ions in the spinels. For a color version of this figure, see www.iste.co.uk/valls/inorganic.zip*

NOTE 3.25.– Spinels differ due to their direct, inverse or intermediate character, despite their identical formula. The sites occupied by metallic ions define their type. In the direct spinels, ions A occupy the tetrahedral sites and ions B occupy the octahedral sites ($A^{Tet}B_2^{Oct}O_4$). In inverse spinels, octahedral sites are occupied half by ions A and half by ions B, with the remaining ions B occupying the tetrahedral sites ($A^{Oct}B^{Oct}B^{Tet}O_4$).

Inverse spinels are characterized by their inversion rate (λ) that is defined by the fraction of atom B in the tetrahedral sites. For a normal spinel, λ is null; for an intermediary spinel, λ ranges between 0 and 0.5 and for an inverse spinel, λ is equal to 0.5. This rate varies with temperature, since thermal agitation facilitates the ion displacement from one site to another.

This phenomenon is not described in detail in this book, but it is worth being mentioned, as it plays an important role in the spinel properties.

Depending on the nature of cations, their number, proportion and distribution in the tetrahedral and octahedral sites, spinels may have a wide variety of electronic and magnetic properties. Spinels have various applications as high-frequency inductors or HF transformers, antenna rods, non-volatile magnetic memories, micro-components, etc. Certain spinels are used as magnetic paint for stealth military aircraft and naval vessels coating.

3.6.2.3. *Calculation of various values of $MgAl_2O_4$ unit cell*

The cubic F structure of $MgAl_2O_4$ spinel with lattice parameter a of 808.0 pm contains eight formula units (Z) in the unit cell. There is an eightfold presence of $MgAl_2O_4$, which corresponds to the presence of eight magnesium ions, 16 aluminum ions and 32 oxygen ions (see Figure 3.91).

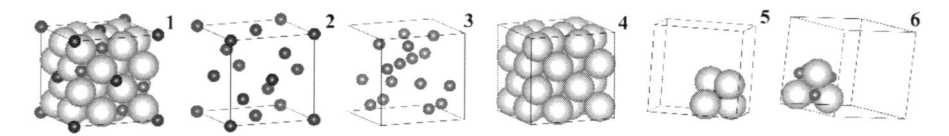

Figure 3.91. *Ion count in spinels. For a color version of this figure, see www.iste.co.uk/valls/inorganic.zip*

The unit cell (1) can be represented only with magnesium ions (2), aluminum ions (3) or finally with oxygen ions (4). The ions can be readily counted: $8 \times 1/8 + 6 \times 1/2 + 4 = 8$ Mg^{2+}, 16 Al^{3+} and 32 O^{2-}.

The eight small cubes, of which four (5) contain four O^{2-} and one Mg^{2+} and four (6) contain four Al^{3+} and four O^{2-} can also be used. The remaining magnesium ions form a cubic F packing that places four atoms in the unit cell.

The radius of Mg^{2+} in coordination 4 is 57 pm and the ionicity of the Mg–O bond is 68%, that of Al^{3+} in coordination 6 is 54 pm and the ionicity of the Al–O bond is 56% and finally, the ionic radius of O^{2-} is 140 pm.

The Al–O distance is 202 pm, and it can be noted that it is above the sum of ionic radii (194 pm), because the size of the unit cell is set by the Mg–O distance and availability is positive (+8 pm) for Al^{3+}.

The Mg–O distance obtained from $a\sqrt{3}/8$ is 174.9 pm, which is less than the sum of ionic radii (197 pm); therefore, this availability is negative (−22.1 pm). The value is partly justified by the value of ionicity as well as by the ion interaction in this structure that has two types of cations. The steric criterion of magnesium is 0.407, which is below the value expected for coordination 4, as for the aluminum cations (0.386), but it has already been noted that the steric criterion is often a secondary criterion for crystals.

Electrostatic valence makes it possible to write 4×1 (Mg) $+ 6 \times 2$ (Al) $= 4 \times 4$ (O).

3.6.2.4. Compounds of $MgAl_2O_4$ type

Similarly to perovskites, the notion of steric criterion makes no sense for spinels, as the two cations play a role in crystal equilibrium and it can be seen that this steric criterion is never simultaneously respected by both cations.

Its calculation shows that only 24% of the compounds present a steric criterion corresponding to coordination and only for one of the two cations (see Table 3.29). The same calculation applied to perovskites leads to 47% correspondence.

The average of steric criteria is close to 0.4 for the two cations, knowing that the limit value between the coordinations 4 and 6 is 0.414. This explains the shift of ions from octahedral sites to tetrahedral sites and vice versa in direct and inverse spinels.

Compounds AB$_2$O$_4$	a	R^+_A/R$^\times$	Avail. A	R^+_B/R$^\times$	Avail. B
CdV$_2$O$_4$ [REU 69]	869.5	0.557	−29.7	0.457	+13.4
MnV$_2$O$_4$ [PLU 62]	852.0	0.471	−21.5	0.457	+9.0
CuAl$_2$O$_4$ [DEL 58]	808.6	0.407	−21.9	0.386	+8.2
MgAl$_2$O$_4$ [ISH 82]	808.0	0.407	−22.1	0.386	+8.0
CuGa$_2$O$_4$ [DEL 58]	839.0	0.407	−15.4	0.443	+7.8
FeV$_2$O$_4$ [REU 69]	845.3	0.450	−20.0	0.457	+7.3
MgV$_2$O$_4$ [RUD 47]	839.4	0.407	−15.3	0.457	+5.8
ZnV$_2$O$_4$ [RUD 47]	839.3	0.429	−18.3	0.457	+5.8
ZnCr$_2$O$_4$ [PAS 30]	828.0	0.429	−20.7	0.443	+5.4
MgGa$_2$O$_4$ [BAF 69]	828.0	0.407	−17.7	0.443	+5.0
MgSc$_2$S$_4$ [PAT 64]	1063.0	0.424	−10.9	0.484	+6.7
CdEr$_2$S$_4$ [TOM 78]	1110.0	0.342	−21.7	0.489	+4.5
FeSc$_2$S$_4$ [PAT 64]	1053.0	0.342	−19.1	0.408	+4.1
MnIn$_2$S$_4$ [LUT 89]	1072.0	0.359	−17.9	0.435	+4.0
MgLu$_2$S$_4$ [PAT 64]	1095.0	0.310	−3.9	0.467	+3.7
MnLu$_2$S$_4$ [PAT 64]	1092.0	0.359	−13.6	0.467	+3.0
CoIn$_2$S$_4$ [LUT 89]	1058.0	0.304	−10.9	0.435	+0.5
FeLu$_2$S$_4$ [PAT 64]	1081.0	0.342	−13.0	0.467	+0.2
MgYb$_2$S$_4$ [PAT 64]	1096.0	0.310	−3.8	0.299	−0.1
MnYb$_2$S$_4$ [PAT 64]	1095.0	0.310	−12.9	0.489	−0.3
CuCr$_2$S$_4$ [PAT 64]	869.5	0.310	+15.6	0.408	−0.5
CuCo$_2$S$_4$ [RIE 73]	946.4	0.310	−36.1	0.337	−2.4
FeYb$_2$S$_4$ [PAT 64]	1084.0	0.359	−12.4	0.489	−3.1

Table 3.29. *Several examples of spinels of oxide and sulfide types*

The review of steric criteria is complex, as two criteria should be considered for each crystal (one for each ion). It can be noted that the one for cation A is always negative, which imposes a certain covalence in its bond.

The hard sphere model can be validated, as the availability of cations B is within tolerance interval (Avail. B), if we admit one covalence for cation A (Avail. A is strongly negative). Only two compounds (MgYb$_2$S$_4$ and MgLu$_2$S$_4$) have the appropriate tolerances for their availabilities.

It can be noted that the compounds with the strongest negative availabilities (FeYb$_2$S$_4$ and CuCo$_2$S$_4$) have also strongly negative anionic availabilities (−12.4 and −36.1 pm) that indicate a significant part of covalence.

The hard sphere model can be validated similarly to perovskites, but the simultaneous presence of several cations may, in some cases, lead to a coexistence of positive availabilities for one cation and negative availabilities for the other.

Ionic and partially covalent crystals have been described with the help of the very simple models of hard spheres and perfect crystal, with equally simple criteria such as ionicity, steric criterion and availability.

The limits of these models have been stated, but most of the compounds are within the defined framework.

Covalent compounds such as graphite or diamond, as well as molecular compounds such as ice or iodine, and many others, have not been discussed as this book does not wish to cover the entire realm of complexity. Rather, its objective is to elicit the curiosity of the reader and motivate the reader to explore the universe of solid matter further.

Appendix

Ionic Radii

All the values of ionic radii used in this book are listed in the below table [HAY 15]. They are proposed for ions with coordination numbers 4, 6, 8 and 12; however, there may be cases where values of certain ions with coordination numbers 2, 3, 5, 10, etc. have not been mentioned (see [HAY 15]), as they are not required. The values of covalent radii (cov) and van der Waals radii (vdW) are also proposed.

ion	coordination				bond		ion	coordination				bond	
	4	6	8	12	cov	vdW		4	6	8	12	cov	vdW
Ag^+	-	115	128	-	136	203	Nb^{3+}	-	72	79		156	207
Al^{3+}	39	54	-	-	124	184	Nd^{3+}	-	98	112	127	188	229
As^{3+}	-	58	-	-	120	185	Ni^{2+}	-	69	-	-	117	184
B^{3+}	-	-	-	-	84	192	Ni^{3+}	-	56	-	-	-	-
Ba^{2+}	-	135	142	161	206	268	O^{2-}	-	140	142	-	64	152
Be^{2+}	27	45	-	-	99	153	OH^-	135	137	-	-	-	-
Bi^{3+}	-	103	117	-	150	207	Os^{4+}	-	63	-	-	136	216
Br^-	-	196	-	-	117	183	P^{3+}	-	-	-	-	109	180
C^{4+}	15	16	-	-	75	170	Pb^{2+}	-	119	129	149	145	202
Ca^{2+}	-	100	112	134	174	231	Pb^{4+}	65	78	94	-	-	-
Cd^{2+}	78	95	110	131	140	230	Pd^{2+}	-	86	-	-	130	202
Ce^{3+}	-	101	114	134	184	235	Pd^{3+}	-	76	-	-	-	-
Cl^-	-	181	-	-	100	175	Pd^{4+}	-	62	-	-	-	-
Cm^{3+}	-	97	-	-	168	245	Pm^{+3}	-	97	109	-	186	236
Cm^{4+}	-	85	95	-	-	-	Pt^{2+}	-	80	-	-	130	209
Co^{2+}	56	65	90	-	118	192	Pt^{4+}	-	63	-	-	-	-
Co^{3+}	-	55	-	-	-	-	Rb^+	-	152	161	172	215	303

Figure A.1. *Values of ionic radii for ions at various coordination numbers, covalent radii and Van der Waals radii (part 1)*

	coordination				bond			coordination				bond	
ion	4	6	8	12	cov	vdW	ion	4	6	8	12	cov	vdW
Cr^{2+}	-	73	-	-	130	189	S^{2-}	-	184	-	-	104	180
Cr^{4+}	41	55	-	-	-	-	Sb^{3+}	-	76	-	-	140	206
Cs^+	-	167	174	188	238	343	Sc^{3+}	-	75	87	-	159	216
Cu^+	60	77	-	-	122	186	Se^{2-}	-	198	-	-	118	190
Cu^{2+}	-	73	-	-	-	-	Si^{4+}	26	40	-	-	114	210
Eu^{2+}	-	117	125	135	183	233	Sn^{2+}	-	-	-	-	140	217
Eu^{3+}	-	95	107	-	-	-	Sn^{4+}	55	69	81	-	-	-
F^-	-	133	-	-	60	147	Sr^{2+}	-	118	126	144	190	249
Fe^{2+}	63	61	92	-	124	194	Ta^{+3}	-	72	-	-	158	217
Fe^{3+}	49	55	92	-			Ta^{+4}	-	68	-	-	-	-
Ga^{3+}	47	62	-	-	123	187	Ta^{+5}	-	64	-	-	-	-
Ge^{2+}	-	73	-	-	120	211	Te^{2-}	-	221	-	-	137	206
Ge^{4+}	39	53	-	-			Te^{4-}	66	97	-	-	-	-
H^-	-	-	-	-	32	110	Te^{6+}	43	56	-	-	-	-
Hg^+	-	119	-	-	132	209	Ti^{2+}	-	86	-	-	148	187
Hg^{2+}	96	102	114	-	-	-	Ti^{3+}	-	67	-	-	-	-
I^-	-	220	-	-	136	198	Ti^{4+}	42	61	74	-	-	-
In^{3+}	62	80	-	-	142	193	Tl^+	-	150	159	170	144	196
Ir^{3+}	-	68	-	-	132	202	Tl^{3+}	75	89	98	-	-	-
K^+	137	138	151	164	200	275	U^{3+}	-	103	-	-	183	240
La^{3+}	-	103	116	136	194	240	U^{4+}	-	89	100	117	-	-
Li^+	59	76	92	-	130	181	U^{6+}	52	73	86	-	-	-
Lu^{3+}	57	86	97	-	174	221	V^{2+}	-	79	-	-	144	179
Mg^{2+}	-	72	89	-	140	173	V^{3+}	-	64	-	-	-	-
Mn^{2+}	66	83	96	-	129	197	W^{4+}	-	66	-	-	150	210
Mn^{4+}	39	53	-	-	-	-	Y^{3+}	-	90	102	-	176	219
Mo^{3+}	-	69	-	-	146	209	Yb^{2+}	-	102	114	-	178	242
Mo^{4+}	-	65	-	-	-	-	Zn^{2+}	60	74	90	-	120	210
Na^+	99	102	118	139	160	227	Zr^{4+}	59	72	84	-	164	186

Figure A.2. *Values of ionic radii of ions at various coordination number, covalent and van der Waals radii (part 2)*

Bibliography

[ACH 05] ACHARY S.N., TYAGI A.K., "Synthesis and characterization of mixed fluorides with PbF_2 and ScF_3", *Powder Diffraction*, vol. 20, pp. 254–258, 2005.

[AIG 94] AIGNER K., ETTMAYER P., LENGAUER W. *et al.*, "Lattice parameters and thermal expansion of $Ti(C_xN_{1-x})$, $Zr(C_xN_{1-x})$, $Hf(C_xN_{1-x})$ and TiN_{1-x} from 298 to 1473 K as investigated by high-temperature X-ray diffraction", *Journal of Alloys Compounds*, vol. 215, pp. 121–126, 1994.

[ALD 62] ALDRED A.T., "Intermediate phases involving scandium", *Transactions of the Metallurgical Society of AIME*, vol. 224, pp. 1082–1083, 1962.

[ALS 25] ALSEN N., "Röntgenographische Untersuchungen der Kristallstrukturen von Magnetkies, Breithauptit, Pentlandit, Millerit und verwandten Verbindungen", *Geologiska Föreningens i Stockholm Förhandlingar*, vol. 47, pp. 19–73, 1925.

[AND 01] ANDREAS L., HERBERT J., STEVE H., "Ordering of nitrogen in nickel nitride Ni_3N determined by neutron diffraction", *Inorganic Chemistry*, vol. 40, pp. 5818–5822, 2001.

[ANG 01] ANGENAULT J., *Symétrie structure: cristallochimie du solide*, Vuibert, Paris, 2001.

[ANG 07] ANGENAULT J., *Chimie des groupes principaux*, Vuibert, Paris, 2007.

[ARN 81] ARNAUD Y., CHEVRETON M., "Etude comparative des composes TiX_2 (X= S, Se, Te). Structures de $TiTe_2$ et TiSeTe", *Journal of Solid State Chemistry*, vol. 39, pp. 230–239, 1981.

[BAF 69] BAFFIER N., HUBER M., "Etude par diffraction des rayons X de spinelles oxyfluores de magnésium et de gallium", *Comptes Rendus Hebdomadaires des Séances de l'Académie des Sciences, Série C, Sciences Chimiques (1966)*, vol. 269, pp. 312–331, 1969.

[BAN 41] BANNISTER F.A., "Osbornite, meteoric titanium nitride", *Mineralogical Magazine (1969)*, vol. 26, pp. 36–44, 1941.

[BAR 21] BARTLETT G., LANGMUIR I., "The crystal structure of the ammonium halides above and below the transition temperature", *Journal of the American Chemical Society*, vol. 43, pp. 84–91, 1921.

[BAR 25] BARTH T., "Die Kristallstruktur von Perowskit und verwandter Verbindungen", *Norsk Geologisk Tidsskrift*, vol. 8, pp. 201–216, 1925.

[BAR 62] BARTLETT N., "Xenon hexafluoroplatinate (V) $Xe^+[PtF_6]$", *Proceedings of the Chemical Society*, London, vol. 6, p. 218, 1962.

[BAT 62] BATES C.H., WHITE W.B., "New high-pressure polymorph of zinc oxide", *Science*, vol. 137, pp. 993–993, 1962.

[BAU 71] BAUR W.H., KHAN A.A., "Rutile-type compounds. VI. SiO_2, GeO_2 and a comparison with other rutile-type", *Structures, Acta Crystallographica Section B (24,1968-38,1982)*, vol. 27, pp. 2133–2139, 1971.

[BEC 25] BECKER K., EBERT F., "Die Kristallstrukturen einiger binärer Carbide und Nitride", *Zeitschrift für Physik*, vol. 31, pp. 268–272, 1925.

[BER 74] BERNIER J.C., CHAUVEL C., KAHN O., "Etude structurale et magnetique des oxydes Perovskites Ba_2NbVO_6 et Sr_2TaVO_6", *Journal of Solid State Chemistry*, vol. 11, pp. 265–271, 1974.

[BER 96] BERAN A., LIBOWITZKY E., ARMBRUSTER T., "A single-crystal infrared spectroscopic and X-ray diffraction study of untwinned San Benito perovskite containing OH groups", *Canadian Mineralogist*, vol. 34, pp. 803–809, 1996.

[BES 82] BESANÇON P., "Détermination des rayons ioniques absolus et des charges ioniques relatives dans les structures cristallines", *Acta Crystallographica Section B*, vol. B38, pp. 2379–2388, 1982.

[BEZ 93] BEZZENBERGER R., SCHOELLHORN R., "Electrochemical synthesis of perovskite type oxobismuthates (III/V) from hydroxide melts", *European Journal of Solid State Inorganic Chemistry*, vol. 30, pp. 435–445, 1993.

[BIH 10] BIHOUIX P., DE GUILLEBON B., *Quel futur pour les métaux?*, EDP Sciences, Paris, 2010.

[BIL 39] BILTZ W., KOECHER A., "Beiträge zur systematischen Verwandtschaftslehre über das System Vanadium/Schwefel", *Zeitschrift für Anorganische und Allgemeine Chemie*, vol. 241, pp. 324–337, 1939.

[BLA 11] BLANC J., La sécurité des approvisionnements stratégiques de la France, Report no. 349 (2010–2011), 10 March 2011.

[BOL 93] BOLZAN A.A., FONG C., KENNEDY B.J. *et al.*, "Powder neutron diffraction study of pyrolusite, beta-MnO_2", *Australian Journal of Chemistry*, vol. 46, pp. 939–944, 1993.

[BOL 97] BOLZAN A.A., FONG C., KENNEDY B.J. *et al.*, "Structural studies of rutile-type metal dioxides", *Acta Crystallographica Section B*, vol. 53, pp. 373–380, 1997.

[BON 64] BONDI A., "van der Waals volumes and radii", *Journal of Physical Chemistry*, vol. 68, pp. 441–451, 1964.

[BON 92] BONNEAU P.R., JARVIS R.F. JR., KANER R.B., "Solid-state metathesis as a quick route to transition-metal mixed dichalcogenides", *Inorganic Chemistry*, vol. 31, pp. 2127–2132, 1992.

[BRO 53] BROUS J., FANKUCHEN I., BANKS E., "Rare earth titanates with a perovskite structure", *Acta Crystallographica Section B (1,1948-23,1967)*, vol. 6, pp. 67–70, 1953.

[BUR 96] BURNS P.C., HAWTHORNE F.C., HOFMEISTER A.M. *et al.*, "A structural phase-transition in K(Mg1-xCux)F3 perovskite Sample: x = 0.000", *Physics and Chemistry of Minerals*, vol. 23, pp. 141–150, 1996.

[BUS 70] BUSING W.R., "An interpretation of the structures of alkaline earth chlorides in terms of interionic forces", *Transactions of the American Crystallographic Association*, vol. 6, pp. 57–72, 1970.

[CHA 70] CHARLTON J.S., CORDEY-HAYES M., HARRIS I.R., "A study of the ^{119}Sn Moessbauer isomer shifts in some platinum-tin and gold-tin alloys", *Journal of the Less-Common Metals*, vol. 20, pp. 105–112, 1970.

[CHE 95] CHEN B.-H., WALKER D., SUARD E. *et al.*, "High pressure synthesis of $NdCuO_3$ perovskites (0<d<0.5)", *Inorganic Chemistry*, vol. 34, pp. 2077–2083, 1995.

[CHE 05] CHEN J., LIU H., MARTIN C.D. et al., "Crystal chemistry of $NaMgF_3$ perovskite and high pressure and temperature Sample: P = 4 GPa, T = 1000 C, high-T polymorph", *American Mineralogist*, vol. 90, pp. 1534–1539, 2005.

[CHE 09] CHENG J., ALONSO J.A., SUARD E. et al., "A new perovskite polytype in the high-pressure sequence of Ba IrO_3", *Journal of the American Chemical Society*, vol. 131, pp. 7461–7469, 2009.

[CHO 68] CHOISNET J., DESCHANVRES A., RAVEAU B., "Substitution du zinc par le cadmium dans le spinelle Zn_2SnO_4. Etude de la répartition des cations dans la solution solide $Zn_{2-2x}Cd_{2x}SnO_4$", *Comptes Rendus Hebdomadaires des Séances de l'Académie des Sciences, Série C, Sciences Chimiques (1966)*, vol. 266, pp. 543–545, 1968.

[COS 83] COSTA M.M.R., DE ALMEIDA M.J.M., "Contribution to the study of the electron distribution in VF_2", *Portugaliae physica*, vol. 14, pp. 71–79, 1983.

[COS 93] COSTA M.M.R., PAIXÃO J.A., DE ALMEIDA M.J.M. et al., "Charge densities of two rutile structures: NiF_2 and CoF_2", *Acta Crystallographica Section B*, vol. 49, pp. 591–599, 1993.

[DE 69] DE HAAN Y.M., "Structure refinements, thermal motion and Madelung constants of cadmium iodide- and cadmium hydroxide-type layer structures", *National Bureau of Standards (U.S.)*, Special Publication, pp. 233–236, 1969.

[DEL 58] DELORME C., "L'asymétrie de l'ion cuivre bivalent dans des combinations du "type NaCl" et du "type Spinelle"", *Bulletin de la Société Française de Minéralogie et de Cristallographie (72,1949-100,1977)*, vol. 81, pp. 79–102, 1958.

[DEL 63] DELVES R.T., LEWIS B., "Zinc blende type Hg Te – Mn Te solid solutions", *Journal of Physics and Chemistry of Solids*, vol. 24, pp. 549–556, 1963.

[DES 96] DESGRANGES L., CALVARIN G., CHEVRIER G., "Interlayer interactions in $M(OH)_2$: a neutron diffraction study of $Mg(OH)_2$", *Acta Crystallographica Section B*, vol. 52, pp. 82–86, 1996.

[DOE 39] DOELL W., KLEMM W., "Über die Struktur einiger Dihalogenide", *Zeitschrift für Anorganische und Allgemeine Chemie*, vol. 241, pp. 239–258, 1939.

[DYU 88] DYUZHEVA T.I., KABALKINA S.S., BENDELIANI N.A., "Compressibility and polymorphism of ReO_3 at pressures up to 30 GPa", *Doklady Akademii Nauk SSSR*, vol. 298, pp. 100–102, 1988.

[EBE 33] EBERT F., WOITINEK H., "Kristallstrukturen von Fluoriden. II. Hg F, Hg F_2, CuF und CuF_2", *Zeitschrift für Anorganische und Allgemeine Chemie*, 1933, vol. 210, pp. 269–272, 1933.

[EHR 48] EHRLICH P., "Structure and synthesis of titanium selenides and tellurides", *Angewandte Chemie* (German Edition), vol. 60, pp. 68–68, 1948.

[EHR 61] EHRLICH P., GUTSCHE W., SEIFERT H.J., "Darstellung und Kristallstruktur von Titandibromid", *Zeitschrift für Anorganische und Allgemeine Chemie*, vol. 312, pp. 80–86, 1961.

[ETO 01] ETO T., ISHIZUKA M., KIKEGAWA T. *et al.*, "Pressure-induced structural phase transition in a ferromagnet Cr Te", *Journal of Alloys Compounds*, vol. 315, pp. 16–21, 2001.

[FAR 91] FARLEY T.W.D., HAYES W., HULL S. *et al.*, "Investigation of thermally induced Li+ ion disorder in Li_2O using neutron diffraction", *Journal of Physics: Condensed Matter*, vol. 3, pp. 4761–4781, 1991.

[FIL 16] FILSO M.O., EIKELAND E., ZHANG J. *et al.*, "Atomic and electronic structure transformations in SnS_2 at high pressures: a joint single crystal X-ray diffraction and DFT study", *Dalton Transactions*, vol. 45, pp. 3798–3805, 2016.

[FON 69] FONTBONNE A., GILLES J.C., "Nouveaux nitrures de tantale. Nitrure et oxynitrures mixtes de tantale et de niobium", *Revue Internationale des Hautes Températures et des Réfractaires*, vol. 6, pp. 181–192, 1969.

[FUM 64] FUMI F.G., TOSI M.P., "Ionic sizes and born repulsive parameters in the NaCL-type alkali halides", *Journal of Physics and Chemistry of Solids*, vol. 25, pp. 257–258, 1964.

[FUR 65] FURUSETH S., KJEKSHUS A., SELTE K., "Redetermined crystal structures of $NiTe_2$, PdTe2, PtS_2, $PtSe_2$, and $PtTe_2$", *Acta Chemica Scandinavica (1-27,1973-42,1988)*, vol. 19, pp. 257–258, 1965.

[GOL 26] GOLDSCHMIDT V.M., "Die Gesetze der Krystallochemie", *Naturwissenschaften*, vol. 14, no. 21, pp. 477–485, 1926.

[GOL 43] GOLDSCHMIDT H.J., RAIT J.R., "Silicates of the Perovskite type of structure", *Nature*, vol. 152, pp. 356–356, 1943.

[GRO 55] GROENEVELD M., WILLEM O.J., "Niggliite, a monotelluride of platinum", *American Mineralogist*, vol. 40, pp. 693–696, 1955.

[GRO 56] GRONVOLD F., ROST E., "On the sulfides, selenides and tellurides of palladium", *Acta Chemica Scandinavica (1-27,1973-42,1988)*, vol. 10, pp. 1620–1634, 1956.

[GRO 60] GROENVOLD F., KJEKSHUS A., HARALDSEN H., "On the sulfides, selenides and tellurides of platinum", *Acta Chemica Scandinavica (1-27,1973-42,1988)*, vol. 14, pp. 1879–1893, 1960.

[HAH 59] HAHN H., NESS P., "Über Subchalkogenidphasen des Titans", *Zeitschrift für Anorganische und Allgemeine Chemie*, vol. 302, pp. 17–36, 1959.

[HAI 01] HAINES J., LEGER J.M., GORELLI F. et al., "X-ray diffraction and theoretical studies of the high-pressure structures and phase transitions in magnesium fluoride", *Physical Review, Series B – Condensed Matter (18,1978)*, vol. 64, pp. 1341101–13411010, 2001.

[HAY 15] HAYNES W.M., *CRC Handbook of Chemistry and Physics*, 96th ed., Boca Raton, 2015.

[HES 97] HESTER J.R., TOMIMOTO K., NOMA H. et al., "Electron density in $YTiO_3$", *Acta Crystallographica Section B*, vol. 53, pp. 739–744, 1997.

[HOE 71] HOEKSTRA H.R., SIEGEL S., GALLAGHER F.X., "Reaction of platinum dioxide with some metal oxides", *Advances in Chemistry Series*, vol. 98, pp. 39–53, 1971.

[HOL 68] HOLLAND H.J., BECK K., "Thermal expansion of zinc telluride from 0 to 460°C", *Journal of Applied Physics*, vol. 39, pp. 3498–3499, 1968.

[HOW 00] HOWARD C.J., KENNEDY B.J., CHAKOUMAKOS B.C., "Neutron powder diffraction study of rhombohedral rare-earth aluminates and the rhombohedral to cubic phase transition", *Journal of Physics: Condensed Matter*, vol. 12, pp. 349–365, 2000.

[HUL 94] HULL S., KEEN D.A., "High-pressure polymorphism of the copper(I) halides: a neutron-diffraction study to 10 GPa Note: P = 0 GPa Note: ZnS structure, sphalerite structure Note: known as CuBr-III", *Physical Review B*, vol. 50, pp. 5868–5885, 1994.

[HUL 99] HULL S., KEEN D.A., "Pressure-induced phase transitions in AgCl, AgBr, and AgI locality: synthetic Sample: P = 0.0 GPa, Phase I", *Physical Review B*, vol. 59, pp. 750–761, 1999.

[HUT 81] HUTTON J., NELMES R.J., "High-resolution studies of cubic Perovskites by elastic neutron diffraction II – $SrTiO_3$, $KMnF_3$, $RbCaF_3$ and $CsPbCl_3$", *Journal of Physics C*, vol. 14, pp. 1713–1736, 1981.

[ISH 82] ISHII M., HIRAISHI J., YAMANAKA T., "Structure and lattice vibrations of Mg-Al spinel solid solution sample: stoichiometric $MgO.Al_2O_3$", *Physics and Chemistry of Minerals*, vol. 8, pp. 64–68, 1982.

[JOB 01] JOBIC S., BREC R., PASTUREL A. *et al.*, "Theoretical study of possible iridium ditelluride phases attainable under high pressure", *Journal of Solid State Chemistry*, vol. 162, pp. 63–68, 2001.

[JOV 03] JOVIC N., ANTIC B., KREMENOVIC A. *et al.*, "Cation ordering and order-disorder phase transition in co-substituted $Li_4Ti_5O_{12}$ spinels", *Physica Status Solidi, Section A: Applied Research*, vol. 198, pp. 18–28, 2003.

[KAJ 03] KAJITANI M., TORII S., MATSUDA M. *et al.*, "Neutron diffraction study on lanthanum gallate perovskite compound series", *Chemistry of Materials (1,1989)*, vol. 15, pp. 3468–3473, 2003.

[KAM 75] KAMAT DALAL V.N., KEER H.V., "Studies on the Ni Sy Te1-y, system", *Journal of the Less-Common Metals*, vol. 40, pp. 145–151, 1975.

[KAN 91] KANZAKI M., STEBBINS J.F., XUE X., "Characterization of quenched high pressure phases in $CaSiO_3$ system by XRD and 29Si NMR", *Geophysical Research Letters*, vol. 18, pp. 463–466, 1991.

[KEN 62] KENNA B.T., "The search for technetium in nature", *Journal of Chemistry Education*, vol. 39, no. 9, p. 436, 1962.

[KHI 77] KHITROVA V.I., BUNDULE M.F., PINSKER Z.G., "An electron-diffraction investigation of titanium dioxide in thin films", *Kristallografiya*, vol. 22, pp. 1253–1258, 1977.

[KIR 90] KIRFEL A., EICHHORN K., "Accurate structure analysis with synchrotron radiation. The electron density in Al_2O_3 and Cu_2O", *Acta Crystallographica Section A*, vol. 46, pp. 271–284, 1990.

[KIS 89] KISI E.H., ELCOMBE M.M., "U parameters for the wurtzite structure of ZnS and ZnO using powder neutron diffraction", *Acta Crystallographica Section C*, vol. 45, pp. 1867–1870, 1989.

[KJE 69] KJEKSHUS A., WALSETH K.P., "On the properties of the Cr1+x Sb, Fe1+x Sb, $Co_{1+x}Sb$, $Ni_{1+x}Sb$, $Pd_{1+x}Sb$, and $Pt_{1-x}Sb$ phases", *Acta Chemica Scandinavica (1-27,1973-42,1988)*, vol. 23, pp. 2621–2630, 1969.

[KNI 09] KNIGHT K.S., "Parameterization of the crystal structures of centerosymmetric zone-boundary-tilted perovskites: an analysis in terms of symmetry-adapted basis-vectors of the cubic aristotype phase", *The Canadian Mineralogist*, vol. 47, pp. 381–400, 2009.

[KNO 61] KNOX K., "Perovskite-like fluorides. I. Structures of $KMnF_3$, $KFeF_3$, $KCoF_3$, $KNiF_3$ and $KZnF_3$. Crystal field effects in the series and in $KCrF_3$ and $KCuF_3$", *Acta Crystallographica (1,1948-23,1967)*, vol. 14, pp. 583–585, 1961.

[KOE 57] KOEHLER W.C., WOLLAN E.O., "Neutron-diffraction study of the magnetic properties of perovskite- like compounds $LaBO_3$", *Journal of Physics and Chemistry of Solids*, vol. 2, pp. 100–106, 1957.

[KOM 57] KOMAREK K.L., WESSELY K., "Die Systeme Nickel – Selen und Kobalt – Nickel-Selen", *Zeitschrift für Anorganische und Allgemeine Chemie*, vol. 289, pp. 203–206, 1957.

[KOO 34] KOOPMANS T., "Über die zuordnung von wellenf unktionen und eigenwerten zu den einzelnen elektronen eines atoms", *Physica*, vol. 1, nos. 1–6, pp. 104–113, 1934.

[KRA 67] KRAUSKOFT K.B., *Introduction of Geochemistry*, McGraw-Hill, New York, 1967.

[KRO 03] KROLL P., MILKO M., "Theoretical investigation of the solid state reaction of silicon nitride and silicon dioxide forming silicon oxynitride (Si_2N_2O) under pressure", *Zeitschrift für Anorganische und Allgemeine Chemie*, vol. 629, pp. 1737–1750, 2003.

[KRU 52] KRUG J., SIEG L., "Die Struktur der Hochtemperatur-Modifikationen des Cu Br und Cu", J, *Zeitschrift für Naturforschung, Teil A. Physikalische Chemie, Kosmophysik (2,1947-32,1977)*, vol. 7, pp. 369–371, 1952.

[KWE 93] KWEI G.H., LAWSON A.C., BILLINGE S.J.L. et al., "Structures of the ferroelectric phases of barium titanate", *Journal of Physical Chemistry*, vol. 97, pp. 2368–2377, 1993.

[LAU 89] LAURIAT J.P., CHEVRIER G., BOUCHERLE J.X., "Space group of U4O9 in the beta phase Sample: F-43m refinement", *Journal of Solid State Chemistry*, vol. 80, pp. 80–93, 1989.

[LE 94] LE CLANCHE M.C., DEPUTIER S., BALLINI Y. et al., "Solid state phase equilibria in the Ni-Ga-Sb system: experimental and calculated determinations", *Journal of Alloys Compounds*, vol. 206, pp. 21–29, 1994.

[LEV 70] LEVY-CLEMENT C., MORGENSTERN BADARAU I., BILLIET Y. et al., "Mise en évidence d'une nouvelle variété structurale de type spinelle lacunaire de l'oxyde mixte $CdSnO_3$", *Comptes Rendus Hebdomadaires des Séances de l'Académie des Sciences, Série C, Sciences Chimiques (1966)*, vol. 270, pp. 1860–1862, 1970.

[LIU 80] LIU L., "The high-pressure phase transformations of PbO_2: an in-situ X-ray diffraction study sample: P = 160 kbar, before laser heat, fluorite structure", *Physics and Chemistry of Minerals*, vol. 6, pp. 187–196, 1980.

[LOD 01] LODZIANA Z., PARLINSKI K., HAFNER J., "*Ab initio* studies of high-pressure transformations in GeO_2", *Physical Review, Series B – Condensed Matter (18,1978)*, vol. 63, pp. 1341061–1341067, 2001.

[LUT 89] LUTZ H.D., JUNG M., "Kationenverteilung und Überstrukturordnung in ternären und quaternären Sulfidspinellen $M^{(II)}(M_2)^{(III)}S$", *Einkristallstrukturuntersuchungen, Zeitschrift für Anorganische und Allgemeine Chemie*, vol. 579, pp. 57–65, 1989.

[MAH 11] MAHMOOD N.B., AL-SHAKARCHI E.K., "Three techniques used to produce BaTiO3 fine powder", *Journal of Modern Physics*, vol. 2, pp. 1420–1428, 2011.

[MAR 25] MARK H., TOLKSDORF S., "Über das Beugungsvermögen der Atome fuer Roentgenstrahlen", *Zeitschrift für Physik*, vol. 33, pp. 681–687,1925.

[MAT 92] MATKOVIC T., MATKOVIC P., "Constitutional study of the Ti, Zr and Hf tellurides", *Metalurgija*, vol. 31, pp. 107–110, 1992.

[MAY 78] MAYER H.W., SCHUBERT K., ELLNER M., "Kristallstruktur von Ir5 Sn7", *Journal of the Less-Common Metals*, vol. 61, pp. 1–7, 1978.

[MEG 46] MEGAW H.D., "Crystal structure of double oxides of the perovskite type", *Proceedings of the Physical Society*, vol. 58, pp. 133–152, 1946.

[MEI 67] MEISALO V., INKINEN J., "An X-ray diffraction analysis of potassium bromide", *Acta Crystallographica*, vol. 22, pp. 58–65, 1967.

[MEL 03] MELNYK G., TREMEL W., "The titanium–iron–antimony ternary system and the crystal and electronic structure of the interstitial compound Ti_5FeSb_2", *Journal of Alloys Compounds*, vol. 349, pp. 164–171, 2003.

[MET 77] METHFESSEL S., SHELTON R.N., SCHILLING J.S.V. et al., "LaAg under hydrostatic pressure: superconductivity and phase transformation", *Solid State Communications*, vol. 24, pp. 659–664, 1977.

[MIC 16] MICHEL F., *Dictionnaire illustré de géologie: initiation aux sciences de la terre*, Belin, Paris, 2016.

[MIT 00] MITCHELL R.H., CHAKHMOURADIAN A.R., WOODWARD P.M., "Crystal chemistry of perovskite-type compounds in the tausonite-loparite series, $(Sr_{1-2x}Na_xLa_x)TiO_3$ Sample: x = 0.00", *Physics and Chemistry of Minerals*, vol. 27, pp. 583–589, 2000.

[MIW 93] MIWA K., FUKUMOTO A., "First-principles calculation of the structural, electronic, and vibrational properties of gallium nitride and aluminum nitride", *Physical Review, Series B – Condensed Matter (18,1978)*, vol. 48, pp. 7897–7902, 1993.

[MOL 58] MOLLER C.K., "The structure of perovskite-like cesium plumbo trihalides", *Nature*, vol. 182, pp. 1436–1436, 1958.

[MOL 65] MOLODYAN I.P., RADAUTSAN S.I., "Solid solutions based on indium antimonide in the indium-antimony-tellurium system", in NASLEDOV D.N. (ed.), *Soviet Research in New Semiconductor Materials*, Consultant Bureau, New York, 1965.

[MOM 11] MOMMA K., IZUMI F., "VESTA 3 for three-dimensional visualization of crystal, volumetric and morphology data", *Journal of Applied Crystallography*, vol. 44, pp. 1272–1276, 2011.

[MOR 87] MORITA N., ENDO T., SATO T. et al., "TiF_2 with fluorite structure - a new compound", *Journal of Materials Science Letters*, vol. 6, pp. 859–861, 1987.

[NAK 05] NAKATSUKA A., OHTAKA O., ARIMA H. et al., "Cubic phase of single-crystal $LaAlO_3$ perovskite synthesized at 4.5 GPa and 1273 K", *Acta Crystallographica Section E*, vol. 61, pp. i148–i150, 2005.

[NAK 15] NAKATSUKA A., ARIMA H., OHTAKA O. et al., "Crystal structure of $SrGeO_3$ in the high-pressure perovskite-type phase", *Acta Crystallographica Section E*, vol. 71, pp. 502–504, 2015.

[NAS 77] NASR EDDINE M., BERTAUT E.F., ROUBIN M. et al., "Etude cristallographique de $Cr_{(1-x)}V_{(x)}N$ a basse temperature", *Acta Crystallographica Section B (24,1968-38,1982)*, vol. 33, pp. 3010–3013, 1977.

[NAT 29] NATTA G., PASSERINI L., "Soluzioni solide, isomorfismo e simmorfismo tra gli ossidi dei metalli bivalenti. – 1. Sistemi: CaO-CdO, CaO-MnO, CaO-CoO, CaO-NiO, CaO-MgO", *Gazzetta Chimica Italiana*, vol. 59, pp. 129–154, 1929.

[NEW 65] NEWLANDS J.A.R., "On the law of octaves", *Chemical News*, vol. 12, p. 83, 1865.

[OFT 27] OFTEDAL I., "Die Gitterkonstanten von Ca O, Ca S, Ca Se, Ca Te.", *Zeitschrift für Physikalische Chemie (Leipzig)*, vol. 128, pp. 135–158, 1927.

[ONO 87] ONODO M., WADA H., "The titanium–sulphur system: structures of $Ti_{6.9}S_9(18H)$ and $Ti_{8.2}S_{11}(33R)$ and the unit cells of 45R and 57R types of titanium sulphide", *Journal of the Less-Common Metals*, vol. 132, pp. 195–207, 1987.

[OTO 01] O'TOOLE N.J., STRELTSOV V.A., "Synchrotron X-ray analysis of the electron density in CoF_2 and ZnF_2", *Acta Crystallographica Section B*, vol. 57, pp. 128–135, 2001.

[OTT 23] OTT H., "Die Raumgitter der Lithiumhalogenide", *Physikalische Zeitschrift*, vol. 24, pp. 209–213, 1923.

[OTT 26] OTT H., "Die Strukturen von MnO, MnS, AgF, NiS, SnI_4, $SrCl_2$, BaF_2, Präzisionsmessungen einiger Alkalihalogenide", *Zeitschrift für Kristallographie, Kristallgeometrie, Kristallphysik, Kristallchemie (144,1977)*, vol. 63, pp. 222–230, 1926.

[PAS 30] PASSERINI L., "Ricerche sugli Spinelle-II. I composti. $CuAl_2O_4$, $MgAl_2O_4$, $MgFe_2O_4$, $ZnAl_2O_4$, $ZnCr_2O_4$, $ZnFe_2O_4$, $MnFe_2O_4$", *Gazzetta Chimica Italiana*, vol. 60, pp. 389–399, 1930.

[PAT 64] PATRIE M., FLAHAUT J., DOMANGE L., "Sur une nouvelle série de spinelles soufres, contenant des terres rares ou du scandium", *Comptes Rendus Hebdomadaires des Séances de l'Académie des Sciences (1884–1965)*, vol. 258, pp. 2585–2586, 1964.

[PAU 29] PAULING L., "The principles determining the structure of complex ionic crystals", *Journal of the American Chemical Society*, vol. 51, no. 4, pp. 1010–1026, 1929.

[PAU 60] PAULING L., *The Nature of the Chemical Bond*, Cornell University Press, 1960.

[PLU 62] PLUMIER R., "Magnétisme – Etude par diffraction des neutrons du compose spinelle MnV_2O_4", *Comptes Rendus Hebdomadaires des Séances de l'Académie des Sciences*, vol. 255, pp. 2244–2246, 1962.

[POT 73] POTOFF A.D., CHAMBERLAND B.L., KATZ L., "A single crystal study of eight-layer barium manganese oxide, $BaMnO_3$", *Journal of Solid State Chemistry*, vol. 8, pp. 234–237, 1973.

[RED 62] REDMAN M.J., STEWARD E.G., "Cobaltous oxide with the zinc blende/wurtzite-type crystal structure", *Nature*, vol. 193, pp. 867–867, 1962.

[REN 72] RENAULT N., BAFFIER N., HUBER M., "Distribution Cationique et Distorsion Cristalline dans les Manganites Spinelles $NiCr_xMn_{2-x}O_4$", *Journal of Solid State Chemistry*, vol. 5, pp. 250–254, 1972.

[REU 69] REUTER B., RIEDEL E., HUG P. et al., "Zur kristallchemie der vanadin$^{(III)}$-spinelle", *Zeitschrift für Anorganische und Allgemeine Chemie*, vol. 369, pp. 306–312, 1969.

[RIE 73] RIEDEL E., HORVATH E., "Spinelle mit substituierten Nichtmetallteilgittern. IV. Röntgenographische Untersuchung der Systeme $CuCr_2(S1_{1-x}Se_x)_4$ und $CuCr_3(Se_{1-x}Te_x)_4$", *Zeitschrift für Anorganische und Allgemeine Chemie*, vol. 399, pp. 219–224, 1973.

[RUD 47] RUDORFF W., REUTER B., "Die struktur der magnesium- und zink-vanadinspinelle", *Zeitschrift für Anorganische und Allgemeine Chemie*, vol. 253, pp. 194–208, 1947.

[RYB 00] RYBCZYNSKI J., RATUSZNA A., WASKOWSKA A. et al., "Investigation of the crystal structure of the hexagonal $RbZnF_3$ perovskite by powder and single crystal X-ray diffraction methods", *Materials Science Forum*, vol. 321, pp. 942–946, 2000.

[SAS 79] SASAKI S., FUJINO K., TAKÉUCHI Y., "X-ray determination of electron-density distributions in oxides, MgO, MnO, CoO, and NiO, and atomic scattering factors of their constituent atom", *Proceedings of the Japan Academy, Series B: Physical and Biological Sciences*, vol. 55, pp. 43–48, 1979.

[SCH 14] SCHREYER M., GUO L., THIRUNAHARI S. et al., "Simultaneous determination of several crystal structures from powder mixtures: the combination of powder X-ray diffraction, band-target entropy minimization and Rietveld methods", *Journal of Applied Crystallography*, vol. 47, pp. 659–667, 2014.

[SEI 02] SEIFERT-LORENZ K., HAFNER J., "Crystalline intermetallic compounds in the K-Te system: the Zintl-Klemm principle revisited", *Physical Review, Series B – Condensed Matter (18,1978)*, vol. 66, pp. 094105-1–094105-15, 2002.

[SEI 04] SEISENBAEVA G.A., SUNDBERG M., NYGREN M. et al., "Thermal decomposition of the methoxide complexes $MoO(OMe)_4$, $Re_4O_6(OMe)_{12}$ and $(Re_{1-x}Mo_x)O_6(OMe)_{12}$ ($0.24 < x < 0.55$)", *Materials Chemistry and Physics*, vol. 87, pp. 142–148, 2004.

[SHA 69] SHANNON R.D., PREWITT C.T., "Effective ionic radii in oxides and fluorides", *Acta Crystallographica Section B*, vol. B25, pp. 925–946, 1969.

[SHE 08] SHEN J.-M., FENG Y.-T., "Formation of flower-like carbon nanosheet aggregations and their electrochemical application", *Journal of Physical Chemistry C*, vol. 112, pp. 13114–13120, 2008.

[SHI 03] SHIRAKI K., TSUCHIYA T., ONO S., "Structural refinements of high-pressure phases in germanium dioxide", *Acta Crystallographica Section B*, vol. 59, pp. 701–708, 2003.

[SLA 64] SLATER J.C., "Atomic radii in crystals", *Journal of Chemical Physics*, vol. 41, no. 10, pp. 3199–3204, 1964.

[SOD 79] SODECK H., MIKLER H., KOMAREK K.L., "Transition metal-chalcogen systems. VI: the zirconium-tellurium phase diagram", *Monatshefte für Chemie (108, 1977)*, vol. 110, pp. 1–8, 1979.

[STE 60] STENBERG E., ASELIUS J., ARONSSON B., "Borides of rhenium and the platinum metals. The crystal structures of Re_7B_3, ReB_3, Rh_7B_3, $RhB_{1.1}$, $IrB_{1.1}$ and PtB", *Acta Chemica Scandinavica (1-27, 1973-42,1988)*, vol. 14, pp. 733–741, 1960.

[SUT 65] SUTTON L.E., *Tables of Interatomic Distances and Configurations in Molecules and Ions*, nos. 11–18, The Chemical Society, London, 1965.

[TAD 90] TADAKI S., "Electrical properties of $NbTe_4$ and $TaTe_4$", *Synthetic metals*, vol. 38, pp. 227–234, 1990.

[TAY 84] TAYLOR D., "Thermal expansion data: I. Binary oxides with the sodium chloride and wurtzite structure, MO", *Transactions and Journal of the British Ceramic Society*, vol. 83, pp. 5–9, 1984.

[TER 92] TERZIEFF P., IPSER H., "A contribution to the ternary phase diagram Co-Sb-Te", *Monatshefte für Chemie und verwandte Teile anderer Wissenschaften (109,1978)*, vol. 123, pp. 35–42, 1992.

[THO 40] THOMASSEN L., "An X-ray investigation of the system Cr_2O_3-NiO", *Journal of the American Chemical Society*, vol. 62, pp. 1134–1135, 1940.

[TOM 78] TOMAS A., SHILO I., GUITTARD M., "Structure cristalline du Spinelle $CdEr_2S_4$", *Materials Research Bulletin*, vol. 13, pp. 857–859, 1978.

[TRO 81] TRONC E., MORET R., "Structure refinement of nonstoichiometric Ti S2 ($Ti_{1.083}S_2$)", *Synthetic Metals*, vol. 4, pp. 113–118, 1981.

[VAI 56] VAINSHTEIN B.K., "Refinement of the structure of the group NH_4 in the structure of ammonium chloride", *Trudy Instituta Kristallografii Akademiya Nauk SSSR*, vol. 12, pp. 18–24, 1956.

[VER 09] VERBRAEKEN M.C., SUARD E., IRVINE J.T.S., "Structural and electrical properties of calcium and strontium hydrides", *Journal of Materials Chemistry*, vol. 19, p. 2766, 2009.

[VIL 59] VILLADSEN J., "Note on the crystal structure of vanadium dichloride", *Acta Chemica Scandinavica (1-27,1973-42,1988)*, vol. 13, pp. 2146–2146, 1959.

[WAL 04] WALKER D., VERMA P.K., CRANSWICK L.M.D. et al., "Halite-sylvite thermoelasticity. Sample: msl416031, T = 25C, P = 0.0 kbar, cell volume = 179.42 Å^3", *American Mineralogist*, vol. 89, pp. 204–210, 2004.

[WAS 23] WASASTJERNA J.A., "On the radii of ions", *AcTa Societatis Scientiarum Fennicae*, vol. 38, pp. 1–25, 1923.

[WEI 62] WEIR C.E., PIERMARINI G.J., "A diamond cell for X-ray diffraction studies at high pressures", *Journal of Research of the National Bureau of Standards, Section A. Physics and Chemistry*, vol. 66, pp. 325–331, 1962.

[WES 31] WESTGREN A., STENBECK S., "Roentgenanalyse der Gold-Zinn-Legierungen", *Zeitschrift fuer Physikalische Chemie, Abteilung B: Chemie der Elementarprozesse, Aufbau der Materie*, vol. 14, pp. 91–96, 1931.

[WYC 31] WYCKOFF R.W.G., *The Structure of Crystals*, 2nd ed., The Chemical Catalog Company, New York, 1931.

[WYC 63a] WYCKOFF R.W.G., "Fluorite structure", in *Crystal Structures*, vol. 1, 2nd ed., Interscience Publishers, New York, 1963.

[WYC 63b] WYCKOFF R.W.G., "Rocksalt structure", in *Crystal Structures*, vol. 1, 2nd ed., Interscience Publishers, New York, 1963.

[YIM 69] YIM W.M., "Solid solutions in the pseudobinary (III-V)-(II-VI) systems and their optical energy gaps", *Journal of Applied Physics*, vol. 40, pp. 2617–2623, 1969.

[YOS 16] YOSHIASA A., NAKATANI T., NAKATSUKA A. et al., "High-temperature single-crystal X-ray diffraction study of tetragonal and cubic perovskite-type $PbTiO_3$ phases", *Acta Crystallographica Section B*, vol. 72, pp. 381–388, 2016.

[YUR 87] YURI S., ANZAI S., OHTA S. et al., "Magnetic susceptibility, electrical resistivity and thermal expansion coefficient of NiAs-type V1-x Crx Se", *Journal of Magnetism and Magnetic Materials*, vol. 70, pp. 215–217, 1987.

[ZAC 26] ZACHARIASEN W.H., "Die Kristallstruktur der Telluride von Zink, Cadmium und Quecksilber", *Norsk Geologisk Tidsskrift*, vol. 8, pp. 302–306, 1926.

[ZHA 99] ZHANG J., "Room temperature compressibilities of MnO and CdO: further examination of the role of cation type in bulk modulus systematics. Sample from Run 1: P = 0.00 GPa at beginning of experiment", *Physics and Chemistry of Minerals*, vol. 26, pp. 644–648, 1999.

[ZHU 62] ZHURAVLEV N.N., SMIRNOVA YE.M., ZHDANOV G.S., "Investigation of ternary solid solutions on the basis of superconducting compounds", *Physics of Metals and Metallography*, vol. 13, pp. 55–61, 1962.

[ZIN 34] ZINTL E., HARDER A., DAUTH B., "Gitterstruktur der Oxide, Sulfide, Selenide und Telluride des Lithiums, Natriums und Kaliums", *Zeitschrift für Elektrochemie und angewandte physikalische Chemie*, vol. 40, pp. 588–593, 1934.

Index

A, B, C

abundance of elements, 19, 20
alloy, 8, 103, 104, 109, 111, 119–123, 159
anionic availability, 156, 204, 210, 211, 214
annealing, 109, 114
anti-cuboctahedron, 66, 100, 200
atmophiles, 20
atomic radius, 25, 27–29, 42
Bravais lattices, 82, 83
cadmium
  chloride, 212
  iodide, 206
CaF_2, 86, 127, 133, 145, 149, 150, 191–196, 216–219
calcium fluoride, 191
$CdCl_2$, 86, 127, 137, 139, 142, 147, 149, 150, 204, 212–216, 218–220
CdI_2, 86, 127, 139, 142, 147, 149, 150, 204–216, 218–220
cesium chloride, 151
chalcogenides, 220
chalcophiles, 21
close-packed packing, 61
compounds of CsCl type, 155–157
congruent melting, 118

coordination
  coordination 2, 138
  coordination 3, 137–139, 142, 200, 202, 212
  coordination 4, 134, 138, 141, 167, 175, 176, 183, 193, 195, 197, 198, 209, 214, 216, 226, 233
  coordination 6, 134, 135, 141–143, 146, 165, 166, 168, 169, 177, 183, 195, 198, 200, 204, 205, 209, 211, 216, 226, 233
  coordination 8, 135, 136, 153–155, 193–195, 197, 216
  coordination 12, 136, 143, 146, 190
  polyhedra, 128, 129, 142–146, 153, 162–165, 173, 174, 179, 180, 186, 187, 192, 193, 200, 201, 207, 208, 212, 213, 222, 223, 230, 231
  polyhedron, 138, 141, 152, 154, 163, 164, 173, 174, 180, 187, 192, 197
crystal faces, 79
crystalline
  class, 83–85, 91
  systems, 84, 85
crystallographic systems, 84, 85

CsCl, 86, 127, 135, 136, 139, 149–152, 155–160, 166, 168, 169, 171, 174, 184, 190, 191, 216, 217, 219
cubic I, 66, 67, 72, 77–80, 82, 92, 102
cuboctahedron, 62, 66, 95, 134, 136, 139, 146, 173, 186, 215, 223, 224

D, E, F

diagonalization, 18, 25, 28, 30, 31
diagram for total miscibility, 113
directions, 29, 59, 75, 81, 86–90, 127, 203, 229
electrical conductivity, 56, 57
electron binding energy, 25, 34–37
electronegativity, 25, 28–31, 33, 34, 37, 45, 47, 51, 110, 131, 132, 140, 148, 153, 156, 219, 220
electrostatic valence, 130, 141, 142, 154, 166, 175, 188, 193, 197, 202, 208, 212, 233
element 43, 8
energy bands, 55
eutectic, 114–117, 119, 121, 122
formular group, 82

G, H, I

geometric criteria, 129
gliding, 87
grain, 23, 108, 109, 114, 115
Guyton Morveau, 5
halides, 22, 155, 166, 176, 182, 194, 198, 209, 214, 219, 220
hexagon, 64, 66, 85, 94, 96, 99, 186, 216, 218
hexagonal close-packed, 64, 66, 67, 70, 72, 73, 99, 100, 102, 178–180, 182, 186, 200, 204, 207, 215
high period packings, 68
homogeneity of the environment, 130, 147

intermetallic compounds, 104, 111, 112, 125, 150, 151, 157, 158, 196, 216
interplanar distances, 74
interstitial sites, 100, 128
ionic
  character, 130, 133, 166, 175, 182, 198
  crystals, 37, 125, 126, 129, 144, 149
  radii, 41, 42, 44, 47, 50, 51, 129, 133, 140, 151, 153, 154, 166, 175, 177, 182, 184, 185, 193, 197, 201, 221, 233
ionization potential, 25, 31–34, 36

L, M, N

Lavoisier, 1–5, 7
law of octaves, 8, 15
lever rule, 107, 108, 114, 116, 117, 120, 122, 123
Li_2O, 86, 127, 139, 141, 143, 146, 149, 150, 196–198, 199, 216–218, 220
liquidus, 114, 116
lithium oxide, 196
lithophiles, 20, 22
mantle, 23
melting temperature, 56, 58, 119, 149, 169
Mendeleev, 1–3, 5, 6, 8, 14–16
metallic bond, 54, 59, 149, 186
metalloids, 9, 11, 18, 30, 31, 36, 37, 58, 131, 132
$MgAl_2O_4$, 127, 149, 227, 229, 230, 232–234
Miller indices, 73–75, 81
Moseley, 1–3, 6, 8, 9, 18
NiAs, 49, 86, 127, 135, 139, 144–147, 149, 150, 185–189, 216, 218, 220

nickel arsenide, 135, 185
noble gases, 10, 13, 14, 16, 17, 20, 27, 28, 32–34
non-close-packed, 60, 61, 66, 67, 73
nucleus, 6, 25–27, 32, 42–45, 48

O, P, R

octahedral site, 101, 111, 164, 166, 186, 231, 232, 233
ordered phases, 112, 114
packing
    density, 66, 68, 71–73, 76, 128, 154, 165, 175, 182, 191, 193
    of A planes, 63, 64
    of atoms, 54, 60
    of N planes, 62
pattern, 85, 95, 96, 152, 155, 162, 173, 174, 190, 191
perfect crystal model, 48, 59, 109, 126
perovskites, 127, 150, 221, 222, 225–228, 233, 235
plane nomenclature, 75
point group, 83, 85–87, 89, 90, 154, 166, 175, 182, 194
polarizability, 37, 44–47
polytypism, 68
poor metals, 10, 11, 18
reading a diagram, 107, 108
refractory, 58
rhombohedral, 83, 85–87, 93–96, 159, 171, 178, 212
    system, 93
rhomboidal dodecahedron, 79
rutile, 22, 137, 141, 143, 149, 199–204, 208

S, T, V, Z

semiconductors, 55, 56, 132, 177
siderophiles, 21
sodium chloride, 22, 159
solid solution, 104, 105, 109–113, 116–118, 120–123
solidus, 105, 106, 114
solubility binary diagram, 105, 106
space group, 83–88, 90, 91, 149, 150, 188, 197, 198, 201, 215, 221, 222, 225–228
spindles, 113
spinels, 22, 127, 150, 227–234
$SrTiO_3$, 127, 134–136, 139, 143, 146, 221–226
surface density, 76–78
technetium, 7, 8
tetrahedral site, 102, 110, 175, 182, 191, 197, 231–233
tetrahedron, 79, 80, 96, 98, 99, 102, 103, 129, 134, 139, 146, 171, 172, 174, 180, 181, 184, 191, 192, 197, 229, 231
TiO_2, 22, 86, 127, 133, 137, 139, 141–143, 149, 150, 195, 199–205, 216, 219, 220
topologically identical, 152, 154, 166, 172, 216, 217, 218
typical binary crystals, 127
various diagrams, 117
ZnS (sphalerite), 171–173, 175–177
ZnS (wurtzite), 178–183

Printed and bound by CPI Group (UK) Ltd, Croydon, CR0 4YY